THE WOODWARD WAY

A HISTORY OF WOODWARD GOVERNOR COMPANY

ROCKFORD, ILLINOIS USA 1870-1995

For permission to reproduce part of this book in any form, contact
Communication Services
Woodward Governor Company
5001 North Second Street
Rockford, IL 61125-7001 USA
Phone: 815/877-7441
Internet E-mail Address: Commserv@wgc.woodward.com

Library of Congress Catalog Card Number: 97-90991
ISBN 0-9660572-0-1

Printed in the United States of America by
Tan Books & Publishers Inc.
Rockford, Illinois

Table of Contents

Acknowledgments

i Introduction

1 The Founder and His Family 1856-1872 CHAPTER 1

15 The Infant Company 1872-1890 CHAPTER 2

29 The Springboard to Modest Success 1890-1902 CHAPTER 3

39 Incorporation to Construction 1902-1910 CHAPTER 4

55 The Impact of World War I 1910-1919 CHAPTER 5

67 The Prosperous 1920s 1920-1929 CHAPTER 6

81 From Depression to Expansion 1930-1939 CHAPTER 7

101 The Impact of World War II 1940-1945 CHAPTER 8

127 The Formation of the Corporate Partnership 1945-1954 CHAPTER 9

147 Domestic and Foreign Expansion 1955-1965 CHAPTER 10

165 The End of an Era But Not of an Idea 1966-1976 CHAPTER 11

181 A Time of Growth 1976-1989 CHAPTER 12

197 Positioning for the Future 1990-1995 CHAPTER 13

209 Notes

Acknowledgments

The history of the History

Assistant to the Chairman Leo C. Powelson started our history project in 1986. He commissioned Bob Hodge, professor of history at Beloit (Wisconsin) College to write it and Archivist Patsy Lundberg of DeKalb, Illinois, to organize a vast accumulation of papers and photos into a company archives for reference. Patsy also was the steady hand who guided us through the many edits and kept us faithful to the facts as evidenced by her carefully documented notes pages. Early on, Designer Karen Miller of Rockford was very helpful to us with suggestions for final format—not easy when we wanted answers for something that hadn't been written yet.

Many long-time members and retirees in Illinois and Colorado shared their memories of the company with Bob. Administrative assistants Jorgiann Long and Marsha Opperman transcribed these tape recorded interviews, as well as Bob's handwritten notes.

Leo, Mark Leum (then president), and Jerry Forberg (then Rockton plant manager) did much of the review and edit. Jean Palombi (chairman's office) added her edit expertise and patiently input draft after draft. In 1989, Frank Walters, who was working on his Ph.D. and teaching English at Northern Illinois University, DeKalb, did a complete editorial rewrite to polish our draft into a finished script.

Rockford's Communication Services Department assisted Leo, Bob, Patsy, and Frank throughout the project: Jim Orlando, historical photos; Ami Barker and Michelle Nason, desktop publishing; Fred Bering, writing; Jim Bollenbach, research; and Nat Jurasek, editing, proofreading, and production details.

When Leo was nearly ready to publish the history, our business world began to change drastically. Faced with low profits, reductions in our workforce, a plant closing, and unstable business conditions, it didn't seem prudent to proceed.

Communication Services inherited the project when Leo retired in 1993 after fifty-two years of service. In 1996, our management decided it was time to publish. Writer Fred Bering penned the last chapter of this volume which carries us through 1995, our 125th year, and introduces the new era of Woodward under the leadership of CEO and President John Halbrook.

Designer Mimi Fritz formatted the next-to-final draft and scanned photos into text. Ami Barker did the subsequent fine-tuning and prepared the files on disk for the printer. In July 1997 we gave the disk to Bill Moore of Tan Books and Publishers. For fourteen months, Bill had patiently advised and encouraged us. We needed that. Fittingly, our company's long-term friend and associate, Ken Yanzy, designed the dust cover.

And, here it is, our eleven-year project brought to fruition at last. Our sincere appreciation goes to other unnamed people who contributed to this volume and to all the dedicated members who lived the story.

Introduction

In 1986, the Woodward Governor Company began a project to record the history of the company and the development of its unique business philosophy, the Corporate Partnership. We knew that the past can easily be forgotten in the rush to join the future, and that, once forgotten, can never be recovered again. It was to preserve the company's past and its traditions, and also to show how these have contributed to the company's success, that we saw as our major challenge. Accordingly, we commissioned the writing of the history of the Woodward Governor Company, now issued in this volume as *The Woodward Way.*

The story begins with Amos Woodward's decision in 1856 to leave his home in Maine and move to Rockford, where he eventually started his own business, building and selling governors for water wheels. The story of Woodward Governor Company continues from that beginning to be a drama of people and their interactions with technology and society. Elmer Woodward, Amos's son, combined a talent for business with skill in engineering to smooth the company's transition into the modern world of manufacturing. Irl Martin, Elmer's son-in-law and the company's third chairman, introduced the Corporate Partnership, an innovative business structure that involved all company members in its business success. Following Irl's death in 1976, Cal Covert assumed the responsibility for overseeing the company's development and manufacturing of a very technical product line, and of administering the Corporate Partnership, whose humanistic roots reach back over 100 years.

As these four would testify, the real strength of the company is found in the men and women who work for it. They are the Woodward Way, in body and spirit. They are bonded to the company by the Corporate Partnership, the application of the Woodward Way to the company's day-to-day operation. The company began the Partnership to recognize the interdependence and common goals of worker and shareholder members. Although the story is told in detail in Chapter Nine, it deserves some mention here, because the Partnership is the backbone of the company's post-World War II method of operation.

Actually, it had its beginnings in observations by Irl in the 1930s, during the Great Depression. As the Depression deepened, Irl watched how Elmer and other company leaders avoided layoffs and severe financial losses by cutting back on working hours. Employees had voted for the reduced hours themselves, preferring lower pay but steady work to no work (and no pay) at all. Irl was struck by the sense of family the entire working membership showed during the crisis, and which it would continue to show during the next, greater crisis of

World War II. After the war, Irl made the Partnership the focus, if not the obsession, of the next two years. Many of the elements which evolved into the Woodward Way, such as an early medical plan, a consistent dress code, air conditioning throughout the plants, member committees, and plant meetings were introduced in the late 1930s. The company headquarters in Rockford, built in 1941, incorporated many of these ideas into its very architecture. By 1947, details of the Partnership had been defined clearly enough for it to be presented to the company membership. That same year it was approved and put into effect. In the decades since, it has been refined here, strengthened there, clarified and augmented in many of its details, but always it has remained faithful to its fundamental humanistic principles of honoring the individual above his mere mechanical worth to the company.

The Woodward Way is more, however, than the history of a company; it is also about the way people live honest, productive lives in the worlds of business and industry. The early chapters tell the stories of Amos and Elmer, their families and fellow workers, their community, and their business acumen and inventive skill. When Irl enters the picture, we meet a man who carries on the traditions of his forebearers and follows through on his vision of how these traditions can be made into the philosophy we know as the Woodward Way. Meanwhile, hundreds of Woodward members have selflessly served the company and community in ways too numerous to relate in this book, although we have tried to incorporate as many of their stores as possible in order to illustrate the company's great human potential. For, above all else, the Woodward Way has always been a story of men and women working together to improve their company, their community, and their lives.

The Founder and His Family

1856-1872

In 1856, the year Amos Walter Woodward left home and family to pursue his dreams in the West, John Greenleaf Whittier, the Quaker poet, sought to express the dilemma facing America when he published these lines in "Maud Muller":

> *For of all sad words of tongue or pen,*
> *The saddest are these: "It might have been!"*

The nation was on the brink of dissolution. The agrarian South was witnessing its way of life come under increasing conflict with what it perceived to be the political encroachments of the industrialized North. In Congress, in newspapers and broadsides, in public forums throughout the land, argument raged over states' rights versus the power of a central federal authority, over slavery versus abolition. One would have expected, with Whittier, something to have been done to ward off disaster. Attempts were made, to be sure, but mostly as ineffective compromises administered by weak political leadership. They did little more than delay the inevitable, which became reality soon enough. In December 1860, South Carolina seceded from the Union, followed quickly by Georgia, North Carolina, and Alabama. Thereafter, argument degenerated from words to weapons, from public halls and editorial columns to the battlefield. The following April, with the Confederate States of America self-proclaimed an independent nation, the South Carolina militia opened fire on the federal munitions garrison at Fort Sumter, in Charleston Harbor.

Amos Walter Woodward, age unknown, but clearly a striking figure.

On a much smaller scale, Whittier's words might also have applied to Amos. Born in 1829 to Amos and Nancy (Muzzy) Woodward, farmers in Winthrop, Maine, the young Amos (called Walter by his family) seemed destined to remain an Easterner all his life. When he was twenty, he went to Lowell, Massachusetts, to serve his machinist apprenticeship for a company that made machinery for the cotton mills. This was the customary first step for a young man wishing to enter the skilled trades. Three years later, he was back in Maine to attend Kents Hill Academy for the academic year. The next year, however, saw him in Worcester, Massachusetts, no longer an apprentice, but a toolmaker for Flagg and Company. There he met and befriended George Emerson, who would later play an important role in helping Amos get his first job in Rockford. But, Rockford was just a faraway dot on the map, and Amos was making a good living in Worcester. His skills as a machinist, by now grown considerable, were of just the kind the industrialized Northeast needed most. By 1856, when he was twenty-seven, there seemed no reason for Amos to think of leaving.

But leave he did, that same fall. It is difficult to tell what was going through his mind as he stepped aboard a train bound for Minneapolis, Minnesota, where a cousin lived.[1] But there was nothing unusual in his decision to leave. Hosts of young men and women, lured by the open spaces and expansive opportunities of the West, were leaving the crowded East behind them. The futures they envisioned were as unlimited as the prairie horizon before them. Its unknown mysteries exerted a powerful influence on those restless souls seeking to prove themselves against the yardstick of uncertain but beckoning promise. Perhaps Amos was just such a man as this—drawn to explore the frontiers of his country as well as the limits of his abilities. This same spirit, tempered by age and responsibility, would later lead him to start his own company, and through the years it would influence the gradual development of a unique corporate philosophy.

Amos traveled by train as far as Rockford and by stagecoach to Beloit, Wisconsin, where Charles Fairbanks, a boyhood friend and neighbor, was living. Fairbanks, who worked as a butcher[2], knew other Easterners who had moved to the area and he introduced one of them, a man named Talcott, to Amos. It was Talcott who persuaded Amos to return to Rockford and who provided him with a letter of introduction to his brother Wait.[3] Wait Talcott was part owner with a local industrialist named Ralph Emerson of Talcott, Emerson and Company, manufacturers of the John H. Manny reaper. The reaper's inventor, John H. Manny, had recently won a court case against Cyrus McCormick, who had accused him of patent infringement. The case was argued in Cincinnati by several well-known lawyers, including Edwin M. Stanton and Abraham Lincoln. The court decided in Manny's favor on January 16, 1856, and the decision was later upheld by the U.S. Supreme Court. Manny, however, died on January 31, 1856, when he was only thirty years old. Manny's patented reapers continued to be manufactured by Talcott and Emerson.[4] Back in Rockford, Amos met up again with George Emerson, who ran the Talcott machine shop. Knowing Amos to be a good machinist, George hired him on the spot.[5] This job was the first of four that would help establish Amos's roots in Rockford.

What did Amos know of Rockford? He would, in time, become one of its leading citizens, but as a new arrival in this small city of 5,000, he suffered the usual handicap of being a stranger in a strange land. In 1834, two years after the Blackhawk War, Germanicus Kent and Thatcher Blake had come from Galena to build a sawmill. They cleared some land, raised their mill, and started local history in motion. By 1836, a few other families had arrived and called their settlement "Midway," since it was equidistant between Galena and Chicago. (It would not be called Rockford until 1852, when it was incorporated as a city.) In the years following Kent and Blake's founding, Rockford grew toward civic prosperity and independence. By 1839, it was already a thriving town of 235, with a post office, regular stagecoach service from Chicago, and at least two churches. Over time, Rockford acquired more churches, a newspaper, a bridge and a dam, a courthouse and jail (a sure sign of civilization), more sawmills, a female seminary (later Rockford College), and a bank. By 1850, its population had swelled to 2,000. In 1852, the Galena and Chicago Union Railroad laid tracks to Rockford, and the first wave of Swedish immigrants began arriving.

The next year, the Rockford Water Power Company built a new dam with a mill race 1,700 feet long. That same year, the city earned mention in the East, when the New York Tribune dubbed Rockford the "Forest City," a name that's been with the city ever since.

When Amos arrived in 1856, Rockford had added a volunteer fire department, a free public school, a telegraph line, a flour mill, a foundry, and gaslights. Cultural and literary events were provided by the Young Men's Association and the churches, who persuaded such luminaries as James Russell Lowell, Ralph Waldo Emerson, and Horace Greeley to visit.

Once Amos was settled, he began the correspondence with his family that would go on for most of his life.[6] Typically, most of what we learn of his early years in Rockford we must infer from the letters he got from his family since he seldom wrote often enough, or in enough detail, to please them. They worried that life in the West would corrupt him. His mother frequently asked about his spiritual and physical welfare. A brother, Franklin, and a sister, Priscilla, regularly implored him to return home. Certainly curiosity motivated their often eloquent letters, but the distance separating Amos from home was psychological as much as it was physical. He was outside the sphere of family influence,

beyond their control. How could one watch over a son or brother so far from home, so out of touch with the center of family authority?

The long, inquisitive arm of the Woodward family reached even from Macon County, Illinois, where a cousin named Maria lived. In no position to persuade Amos to move home, she tried a different theme. Why hadn't Amos, now twenty-eight, married? In a letter dated May 31, 1857, she reproached him for his indifference to settling down with a wife:

> *You have lost one noble Yankee girl by your negligence. Now you need not blame me for I tried hard to make the bargain, but it was your fault. Take a lesson from this and warning. I shan't recommend you again. My conscience forbids it for I have come to the conclusion that you are a poor miserable shiftless fellow. Now cousin Walter [Amos] why will not you try to reform?*[7]

Perhaps she believed the business terminology, combined with the blunt hammer of rhetoric, would turn this shiftless fellow into a model of moral, married rectitude. For indeed she did, contrary to her stated promise, intervene once more in December, announcing that she had a "friend":

> *just the lady for you...speak quick for she is in the market and [there will be] lots of sharp bidding.*[8]

Amos wasn't buying. Not a wife, anyway. He declined to enter the bidding and this was the last he would hear from cousin Maria. Not everyone was so anxious to save Amos from himself. He had another brother, Henry, who had once confessed to "a little western fever" himself.[9] Writing in June 1857, Henry suggested that Amos leave Rockford for Kansas, where a mutual friend planned to build a city on the west side of the Missouri River opposite St. Joseph. Henry even offered financial aid to Amos to start a machine shop and further hinted that Amos could expect free land from the company attempting to start the town.[10]

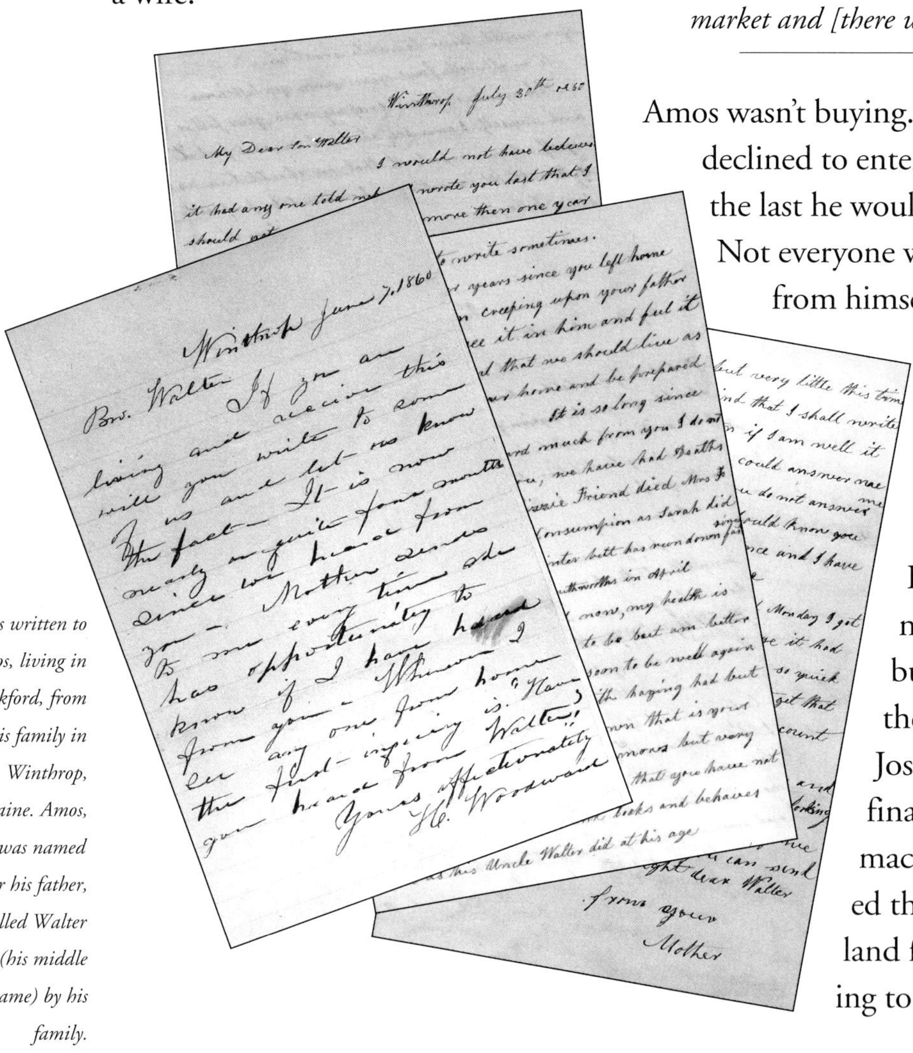

Winthrop June 7. 1860
Bro. Walter If you are living and receive this will you write to some of us and let us know the fact — It is now nearly four months since we have heard from you — Mother desires me every time she has opportunity to know if I have heard from you — When any one from home see me the first inquiry is: "Have you heard from Walter?"
Yours affectionately
H. Woodward

Letters written to Amos, living in Rockford, from his family in Winthrop, Maine. Amos, who was named after his father, was called Walter (his middle name) by his family.

Amos, however, believing that prospects were better in Rockford, declined Henry's advice and money. In the long run, he would be proven correct, but not before going through some hard times which began when Amos lost his job with Talcott during the Panic of 1857. After thirteen years of industrial growth, overexpansion in land and railroads had glutted the economy with speculation and credit, leading to the nation's worst depression in its brief seventy-year history. We get a glimpse into the economic difficulties to come in a letter Henry wrote to Amos in July 1857, before the start of the Panic:

People here begin to think that a reaction must come soon in the business of the West—that there is too much speculation. But still the great work of settling must still go on—though perhaps not so rapidly as of late.[11]

Single and out of work, Amos had the time, the inclination, and apparently the money to leave Rockford and try his luck elsewhere. In the early fall of 1857, he finally made it to Minnesota, his original destination. We know very little about this trip, except that he bought a parcel of land on credit, returned to Rockford, found a part-time job, and excited his family's curiosity once more in a vague letter summarizing his latest doings in the West. His mother, as always, asked about his welfare, but also expressed concern about "this great commotion (the Panic) in money matters."[12] But it took a letter Amos had written to another Maine friend, Sarah Southworth, for his mother to learn that he had found new employment in Rockford. "I am very glad," she wrote, "you could find something to do. If you can earn your board that is doing well to what it would be to do nothing."[13]

Henry also got into the act. Obviously feeling the tension between great national events and his own ambitions, he wrote Amos a letter that begins with some unsolicited advice.

You did not give us much of an account of your trip up to Minnesota... You did not write how long you were gone, how big a piece of land you bought. And, by the way, I am afraid your speculation there will not prove so profitable as it bid fair to.

Speculation, profitable or not, Henry wasn't deterred from renewing his own wishes:

I still have a very strong inclination to the West—Don't you want to go up into Minnesota in the spring and locate?[14]

About this time we begin to see another facet of Amos's character emerge: his deepening interest in religion. No doubt always a religious man, an awakening seems to have occurred in him that would remain the rest of his life. Perhaps the Panic, which had recharged the religious consciousness of the American people, had something to do with it. By the beginning of 1858, the depression was in full force, and widespread and unprecedented unemployment had hit the country's work force.

But the country had not so much rediscovered religion, which it had not really lost, as the comfort religion could provide during hard times. It showed in Rockford where the churches thrived. Daily union prayer meetings were attended by "throngs of worshippers," and people by the hundreds joined the established churches. Amos was one of them. He was hard set upon by the depression, out of a full-time job, worried about paying for his land in Minnesota, and influenced by a deeply religious mother. A logical candidate, therefore, for church membership, he joined the First Baptist Church on May 8, 1858.[15]

Amos was among 140 souls who joined First Baptist that year, although the blessing of expansion was clouded by a few backsliders. "We are pained to say," wrote the church secretary, "that a few of the number give no evidence of true piety and seem to be living entirely for this world."[16]

Amos clearly was not among these; he would, in fact, go on to contribute a long, distinguished career to the church as an officer and elder. But he understood that church membership involved a social as well as a religious life. Joining signified his intention to establish roots in the community, presumably including, someday, a wife and children. But he also believed the church's tenets against "fashionable worldly amusements [that] afford higher pleasure than the service of God," amusements such as dancing, playing cards, and drinking liquor.[17] Indeed, Amos was a lifelong teetotaler who found many like-minded individuals in Rockford. Although a statewide attempt at prohibition had failed in 1855, Rockford had voted ten to one in its favor. For several years thereafter, no liquor licenses were granted within the city limits.[18]

When Nancy learned that her son had joined First Baptist, she was "greatly rejoiced that [Amos] had taken a decided stand and publicly dedicated [himself] to God." She could, therefore, afford to wax philosophical about his financial misfortunes, noting that "our disappointments and misfortunes are for our own good…I trust yours have been a blessing to you."[19]

Blessing or no, Amos also had to support himself, and his prospects that spring and summer of 1858 were dismal, and getting worse. He requested money that his brothers owed him, but they had none to send. He was losing his property in Minnesota. He couldn't borrow money because no bank or private lender would take his note. He had also become ill. Learning of these misfortunes, his family renewed its campaign to get him home and marry a nice Maine girl.[20] By all rights, he should have listened, but things were no better in Maine. Henry, for instance, had lost his boot business and was "living in uncertainty—the worst kind of living there is."[21] But spring is more than a season of new growth; it is also a renewal of hope. And Amos experienced this renewal in the two forms young men most commonly find them: he found work, and he found a bride.

In the spring of 1859, Amos was hired to help build a bridge over the Rock River for a new railroad laying tracks between Rockford and Kenosha, Wisconsin.[22] The economy was on the rebound, and he was able to send money his brother, Franklin, had requested. Amos even hinted to his mother that he was thinking of visiting home in the fall. His family, naturally, hoped that after three years in the West, the prodigal son might come home to stay.[23]

But that summer Amos got a better job at Manny, Blinn & Company as a machinist. This was a new company founded by F. H. Manny to build the John H. Manny reaper. Henry, upon learning that Amos had a full-time job, advised him to make "yourself so useful to them that they cannot get along without you." This Amos apparently did, for by that fall Amos had been promoted to foreman.[24]

Amos had a personal life in Rockford also, which he was customarily reluctant to talk about no matter how hard his family prodded him for details. His mother refused to accept his claims that all his evenings were filled with singing at the church, participating in prayer meetings, socializing at the Ladies Social Society, and attending the Sons of Temperance meetings—all of which he was, but that was also beside the point. What was he doing when he wasn't working or out in public? His sister Priscilla thought she knew. "You must be quite a ladies' man," she wrote him in October 1859, "with some attraction that keeps you where you are." She didn't know how right she was. Henry, for his part, was onto something, too. "Indeed," he wrote in January 1860, "we have been obliged to gather light from your letters somewhat as they learn history from Egyptian Hieroglyphics."[25] As if to implicate, but only subtly, Amos for trying to pull the wool over the family's eyes, Henry, in a letter that April, broadly hinted at his suspicions:

Mary Sealy Woodward, Amos's wife, with their baby son Elmer, who was born October 9, 1862. Elmer would join his father's young company in 1877 and would later succeed him as the company's second president.

Remember that it is the thousand little things of everyday life that will interest [us] such as what you had for breakfast—the color of your boarding mistress' eyes and hair, whether she is a scold or a slut…[26]

It turned out that one of these one thousand "things," and not so little in importance, was Amos's courtship of a young Rockford schoolteacher named Mary Sealy, known to her family by the diminutive "Mate." He had met her sometime in 1858 or early 1859. By 1860, they seem already to have set their marriage plans in motion, for Amos had purchased a lot and started to build a house on it. Finally, in May 1861, Amos confirmed what everyone thought they knew: he and Mate were betrothed.[27] On December 5, they were married at First Baptist.

Young Elmer Woodward

The folks in Maine were overjoyed by the news, and curious about Mate. His mother wanted to know if she was "a professor of religion." Henry joked that he would write the future Mrs. Woodward to "inquire after the whereabouts of her delinquent husband."[28] It would be a long and happy marriage, lasting over fifty-two years and blessed with four children. Elmer, who would play a crucial role in the fledgling company Amos would start the next decade, was born October 9, 1862. Franklin was born in 1867, Cora in 1869, and Minnie in 1874.

When the war broke out in April 1861, Amos was not so preoccupied with his personal life that he could not take an active interest in the events going on. Brought up in a Yankee environment, he disapproved of the Southern secession which had begun the previous December. Only the month before, to express his disapproval, he had voted in his first national election for Abraham Lincoln, an action Henry heartily approved.[29] But the war would prove to be much closer to Amos and Mate than any partisan feelings might indicate.

They had named their first-born after a Civil War hero, one Colonel Elmer Ephraim Ellsworth, who was the first commissioned officer to be killed in the war. How he died stands as a tragic symbol of the nature of war, this one especially. He had personally torn the Confederate flag down from the roof of the Marshall House Hotel in Alexandria, Virginia. For this act of defiance, he was killed by the hotel's proprietor on May 24, 1861. His death was a tragedy especially felt in Rockford, because Ellsworth had left behind a Rockford fiancee. He had other ties to Rockford as well. In 1858, he had come to Rockford to introduce the Zouave System of close-order drill to the Rockford City Grays, a cadet corps and social organization. This method of close-order drill became so popular, that most large cities organized their own Zouave Corps. As a result, Ellsworth became a nationally-known figure. He also was such a close friend and supporter of President Lincoln that his funeral was held in the East Room of the White House. The frenzy over his death was heightened by the distribution of photographs of the officer with the words "Remember Ellsworth" as an epithet for revenge.[30]

The war became a personal tragedy for Mate herself when all three of her younger brothers, Robert, George, and Charles, joined the Union Army. Robert and George would survive, but Charles was captured at the Battle of

Chickamauga in Georgia in September 1863, and later died in the Andersonville prison camp.[31]

Others in Amos's family were affected by the war also, but none, perhaps, more angrily than his sister-in-law (Franklin's wife), Prudentia. In 1867, Jefferson Davis, former president of the Confederate States, was granted bail. Upon hearing this news, Prudentia fired off this strong reaction in a letter to Amos:

I cannot close without speaking of what is uppermost in my mind and was all day yesterday—The news came night before last of the liberation and bailing out of Jeff Davis. Tis the last drop in the full cup of indignities that the country has suffered the last few years. I was not surprised that he was liberated, but I was surprised that Horace Greeley should head the list to give him bail and then go up and shake hands with that Prince of Criminals—Would that his hand might be paralyzed henceforth that he might never write again—I am ashamed of my country. It would be a greater honor to be a Feeje Islander just now. I do love the old Flag yet, but what else is left to us? Men have failed—No law to try a traitor! They had better send a few old women and make some laws! I did hope some soldier would shoot Jeff, but that would not save the dignity of the nation.[32]

Prudentia's wrath shows how difficult it would be to make the nation whole again in body and spirit.

Thirty-two years old when the war began, Amos was past the prime age for soldiering, so he continued working for F. H. Manny, then in the process of establishing an extensive reaper and mower company using the original John H. Manny patents. This large business, capitalized at $60,000 and employing over 100 men, was located on Race Street in what was called the Water Power District, an area on either side of the mill race, on the west bank of the Rock River. Water was supplied to the race by the dam built across the river in 1853. Numerous shops were located along the race, using its water to turn water wheels which powered various kinds of machinery. The Manny facility included a main stone building and three other buildings—one used as a pattern shop and sickle manufacturing facility, another as a blacksmith shop, and the third as a machine storage garage. The Manny Company, one of the largest firms along the Water Power, made farm machinery of all kinds, ultimately emphasizing the Manny Broadcast Seeder and Cultivator.[33]

This building located on Mill Street in the Water Power District, was the A. W. Woodward Company. It included a general pattern and machine shop, as well as production and manufacturing space for governors.

The N. C. Thompson Company's Reaper Works in the Water Power district, dated 1857. Amos worked here in the 1860s. Thompson went out of business in the mid-1880s. From 1893 until 1910, the Woodward Governor Company was located in this building.

Sometime before the war ended, Amos found a new job working for the N. C. Thompson Company, owned by Norman Cornelius Thompson, another early industrialist with a prominent place in local history. He was a native of Georgia who had attended Yale University. During his junior year, his father's store and home burned. This disaster directed his life away from college, and in 1857 he relocated in Rockford. In 1859, he began building a large manufacturing company in the Water Power District. Eventually he would require over $250,000 in capital to conduct this business. During the 1860s, the works covered several acres and employed a work force numbering between 170 and 200 men. The main building was a large machine shop 100 by 70 feet. It was constructed of stone and was four stories high. The various departments, such as the foundry, the paint shop, the blacksmith shop, and the storage sheds, were easily accessible to each other over the private bridge across the race.[34]

This thriving business manufactured plows, cultivators, tooth harrows, and corn planters, but it centered around a contract that Thompson had signed with John Pells Manny in 1860 to make Manny's patented combined reaper and mower on a royalty basis. Thompson had exclusive rights for sales west of the Mississippi River. The Manny machine quickly became popular because it was lightweight, easy to make, and worked equally well as a reaper or mower. For a number of years, the average output was 4,000 machines a year. With each machine selling at about $150 retail, there was profit enough for Thompson and Manny to receive $15 for each machine sold.[35]

By 1860's standards, the workers at the Thompson plant were well paid. The national average for unskilled factory workers in 1860 was about $3 per week for women and $4 per week for men. Skilled mechanics, on the other hand, averaged about $2 a day. Most people worked an eleven-hour day six days a week; a skilled mechanic might make as much as $12 a week. At the Thompson plant, most of the people worked on a piecework system where, based on their ability, they could earn $3 to $7 a day.[36] What Amos averaged is impossible to know, but he was a skilled and inventive machinist, and it would seem likely that he was paid above average wages. Amos at least told his mother that he was doing well, because she replied that she was "pleased that you are prospering so well in your business." Then, to show what else was on her mind besides his economic prosperity, she wrote hoping that "you are also prospering in your spiritual interest."[37] He apparently took his mother's admonition to heart, because in time he moved from active personal involvement to leadership roles at the First Baptist Church. This may have been due in part to Amos's and Mate's attempts to help bring harmony to their church, which was wracked by occasional dissension between and among parishioners and ministers during the 1860s.[38]

For Amos Woodward and his growing family, much would change in the latter 1860s. Most notable was a trip East in the summer of 1865. With Mate and Elmer, who was not quite three, he visited some of Mate's relatives in Geneva, New York (halfway between Rochester and Syracuse), and probably visited his family in Maine. Also, it appears that the Woodwards, to supplement Amos's income, had taken on boarders in 1865 or before. We learn of this from Henry's wife, Sarah, who had written in 1866: "I am very glad to learn you have got to housekeeping again. I think you must feel much happier in your own home then you can be boarding."[39] The boarders would have been a necessary inconvenience, but on September 1, 1865, he had finished paying off his $620 home mortgage with a final payment of $220.[40] Moreover, with Mate pregnant with Elmer, the family would need more space.

Amos's ties with his siblings continued during the late 1860s mainly through correspondence, but the death of his father, in June 1868, would require a solitary trip home. Before this, Amos had to absorb another tragedy: the death, in January 1868, of his childhood friend Charles Fairbanks.[41] Fairbanks had sought his fortune in Beloit, but had never risen above clerk or butcher when he decided to return to Maine.

The death of Amos's father came at the end of a long illness. In early July, immediately after receiving the news, Amos went by rail to Winthrop but arrived too late for the funeral. He was anguished over the death and his inability to be there, and he mused philosophically about the meaning of it all when he wrote Mate from Winthrop:

> *There seems a vacancy that I cannot describe. His hat hangs in the accustomed place and in imagination I can almost see him sitting in the corner or passing out and in, but he is gone to that land where we must all go sooner or later.*[42]

It is a poignant letter, uncharacteristic of Amos for its intensity of feeling and sense of loss, but it shows how close he and Mate had grown. He remained with his family in Maine for over two weeks to comfort his mother and help settle his father's estate. He was not used to being away from his wife and two young boys, and Mate missed him dearly. After telling him that his employer had

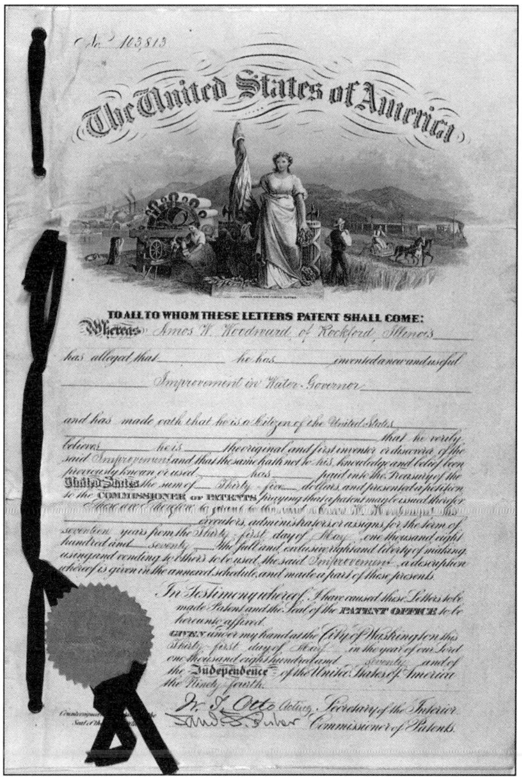
No. 103,813

The United States of America

TO ALL TO WHOM THESE LETTERS PATENT SHALL COME:

Whereas Amos W. Woodward of Rockford, Illinois has alleged that he has invented a new and useful Improvement in Water Governor and has made oath that he is a Citizen of the United States that he verily believes he is the original and first inventor or discoverer of the said Improvement and that the same hath not to his knowledge and belief been previously known or used; has paid into the Treasury of the United States the sum of Thirty five dollars, and presented a petition to the Commissioner of Patents praying that a patent may be issued therefor. [illegible] executors, administrators or assigns for the term of seventeen years from the thirty first day of May, one thousand eight hundred and seventy the full and exclusive right and liberty of making, using and vending to others to be used, the said Improvement, a description whereof is given in the annexed schedule, and made a part of these presents.

In Testimony whereof I have caused these Letters to be made Patent and the Seal of the Patent Office to be hereunto affixed.

Given under my hand at the City of Washington this thirty first day of May in the year of our Lord one thousand eight hundred and seventy and of the Independence of the United States of America the Ninety fourth.

W. T. Otto Acting Secretary of the Interior

Commissioner of Patents

Amos Woodward's first patent, number 103,913, granted May 31, 1870, for "Improvement in Water-Governor." This patent was for the governor Amos built while he was still working for N. C. Thompson Company. Amos built 13 of these in 1871.

cancelled the usual summer shutdown because the reaper business was booming, she called on God to "bless you and keep you from harm and bring you safe home."[43]

About this time, soon after Amos's return from Maine, begins one of the most productive periods of his life, when he would direct his full energies toward water wheel governor technology and the business he would create to build water wheel governors. Water wheels were an especially important technology for the times. They had been in use for generations and were the first widespread source of power in America. Much of the country's industrial development, in fact, can be credited to water power, and on the ability of technology to harness this power for useful ends. This is where the water wheel governor fits in. Its primary purpose was to regulate the flow of water that ran to the wheel from a stream or river. The wheel then provided the motive power for machinery in a factory or mill. Hence the designation of water wheel governors as "prime mover controls." A governor was usually attached to a gate, which it would open or close in response to the power needs of the factory. When the gate was opened wider, more water would flow over the wheel, increasing its rotational speed and thereby feeding more power to the factory.

Amos, because of his experiences in Lowell and Worcester, and now in Rockford, was intimately knowledgeable about water wheel governors, and increasingly dissatisfied with the present state of their technology. Governors were big, cumbersome, slow-acting devices. When factory machines were stopped or started, or whenever water pressure changed due to fluctuations in the race, the governor speed would have to be changed to meet the new power demands. This change was done by hand, and often it took too long for the factory to run efficiently. The governors Amos had to contend with on a daily basis were typical examples. The three "new and splendid" governors Thompson had acquired were, in Amos's judgment, adequate but hardly distinguished representatives of the breed. Next door, at the Dyson Woolen Mill, a new governor, a "huge, ungainly thing," required ten minutes to travel its entire range to open and close the gate. When Amos examined this governor, his attention was caught by a startling fact: in almost every particular, the Dyson governor was no different, and no better, than the governors he'd seen in Lowell twenty years earlier. Like all curious, creative men, Amos asked a simple question with profound implications: shouldn't it be possible to build a better, faster, more efficient governor?[44]

A noncompensating governor of the kind Amos built in 1870. The governor pictured here was installed at a flour mill owned by David Samuels of Marseilles, Illinois. A letter from Mr. Samuels, praising the governor's continuous performance for over thirty years with only one $15.00 overhaul, was featured in a Woodward publication of the early 1900s.

While still employed at Thompson, Amos worked on his own time to find an answer. Finally he made a wooden model, and when Thompson closed down for its summer recess in 1869, Amos was ready to test his invention. On September 1, Amos and his friend and co-worker William McGregor attached the new governor to one of the wheels at the Thompson plant. To their delight, the governor met all their hopes.[45]

The next step for Amos was to obtain a patent on his improved governor. He applied for one on January 20, 1870, and it was granted on May 31, 1870, as Patent Number 103,813, "Improvement in Water-Governor." His new governor improved the speed range of opening or closing the gates from ten minutes to five minutes: a major improvement over other governors. Amos, while still working for Thompson, built thirteen more governors in 1871, one of which was installed in the Dyson Woolen Mill where it remained in operation for over forty years. Amos, at this time, was a tinkerer, but of that rare kind with an eye toward the future. Possessing the curiosity of the experimenter, Amos made the jump from security to the risks of innovation and speculation. Toward the end of 1872, he left the Thompson Company to devote full-time to his governor business. Look upon this decision as another departure to the West, this time not across the open spaces of a country, but across the landscape of the imagination. With this decision, the Woodward Way began.

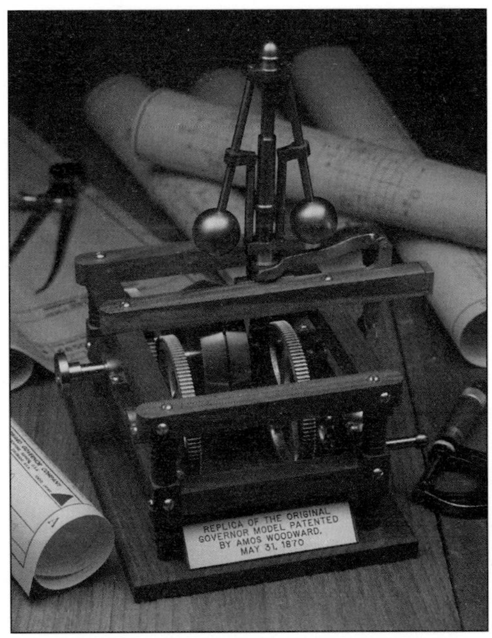

This is a replica of the original governor model patented by Amos Woodward, May 31, 1870.

The Infant Company

1872-1890

With his patent secure, Amos began to build his own governors while still working for N. C. Thompson Company. In 1871, he produced thirteen.[1] There is a company legend that some of these early governors worked badly, but that he had made so many, and had invested so much in them, he was almost forced to improve them rather than drop the whole idea. And improve them he did, so well that by 1872 demand for his product had increased, and so had his desire to launch his own company. Thus, in 1872 when he left Thompson, it was to begin his own business in rented quarters near the woolen mill in the Water Power District.[2] This was a brave undertaking for a forty-three-year-old man with a dependent family, including three young children. He gave up an excellent job on the chance that sales of his governor would give him personal satisfaction and financial independence. The latter, however, despite a frugal life, would not come easily or rapidly to Amos. He had many competitors, and already during the 1870s over a dozen other inventors had received patents for a variety of governors and governor improvements.[3]

In fact, technological advance was everywhere on the map in 1870s America. One expert, a British physicist named John Tyndall, had said, "Mechanical ingenuity engages a greater number of minds in the United States than in any other nation in the world."[4] Tyndall was describing a revolution that was all but leaving the water wheel governor behind. A brief review of this revolution is helpful to an understanding of Amos's bold decision, for he was entering into an area of technology which was about to be replaced by an entirely new generation of machines that produced power.

In 1869, the year Amos built his first governor, water provided almost half the total horsepower used in the United States: 1,130,431 out of 2,346,142. By 1919, however, total horsepower would increase to 29,504,792, whereas that supplied by water barely nudged upward to 1,765,263. Even more revealing are the figures which show the actual rate of decline in the country's use of water power relative to other uses. In 1879, when Amos's business was seven years old, water-generated power had dropped from almost half to 36 percent. Ten years later, it was down to 21 percent, then to 14 percent in 1899, to less than 10 percent in 1909, and finally to less than 6 percent in 1919, the year Amos died.[5] These figures show that any company dependent on water power governing could only hope for a modest (if that) increase in business, unless it branched out into other forms of prime mover controls. Indeed, no matter how many devices Amos and Elmer would invent in the future (and there were many), the company would remain small as long as it specialized in making governors for water wheels and hydro-turbines.

Water wheels, of course, had been used by the earliest civilizations; they were among the human race's first mechanical inventions. Their greatest benefit was that they converted the natural power of flowing water into rotary motion, which facilitated the ability to accomplish work. In colonial times, and up to the early 1840s, water wheels furnished most of the motive power in America. An improvement, the water or hydraulic turbine, occurred at this time which gradually replaced the water wheel. The water turbine didn't make Amos's governor obsolete, since the water wheel governor was easily adaptable to the characteristics of the water turbine. But also at this time, steam engines were on the verge of overtaking water wheels and turbines. And Amos's governor was not as easily adaptable to steam.

The steam engine, though more complex, revealed the limitations of water power technology. The water wheel was obviously limited to industrial sites where an abundant supply of flowing water was available. The water wheel transferred power through an arrangement of wooden or metal shafts and cogwheels. These line shaft systems, though cumbersome and somewhat inefficient by today's standards, were typical of that era. Much of the progress later characterized as the Industrial Revolution had been powered by a configuration consisting of a water wheel and line shaft.

Steam engines quickly became the prime mover of choice by industrial America. For one thing, they carried their water supply with them. For another, steam, under pressure, provided more energy per volume than moving water. Of course, steam engines didn't start out this way. They, too, had to go through a period of improvements before coming into their own. Used successfully in railroads as early as the 1830s, their use in factories lagged because of high purchase cost, a tendency to break down, and the lack of skilled repair technicians to fix and maintain them. By 1840, steam was still five to six times more expensive in annual costs per horsepower than water power. Immediately after the Civil War, however, improvements, especially in metalworking, lowered their initial cost and greatly improved reliability and efficiency. By the 1870s, they had replaced water as the nation's chief source of power.[6]

Other sources of power were also competing with water and steam. In the early 1880s, Thomas Edison had built the first electrical power generating plant. A decade later, the internal combustion gasoline and diesel engines further increased power applications. More than steam or electricity, their invention translated directly into advantages for the common man. They packed relatively large amounts of power—early automobile engines, for example, produced as much as twenty-five horsepower—into a compact, mobile power plant. And what changes they brought to industry were mirrored in society. Because of the automobile, people could roam, privately and at will, wherever there were roads. Put all these new forms of power use together, and they touched off a coal and petroleum revolution that would far surpass, though never eliminate, the importance of water wheels and water turbines. Eventually, of course, hydroelectric generating plants, gas and steam turbines, and diesel engine manufacturers would all become major customers of A. W. Woodward's incorporated successor, today's Woodward Governor Company.

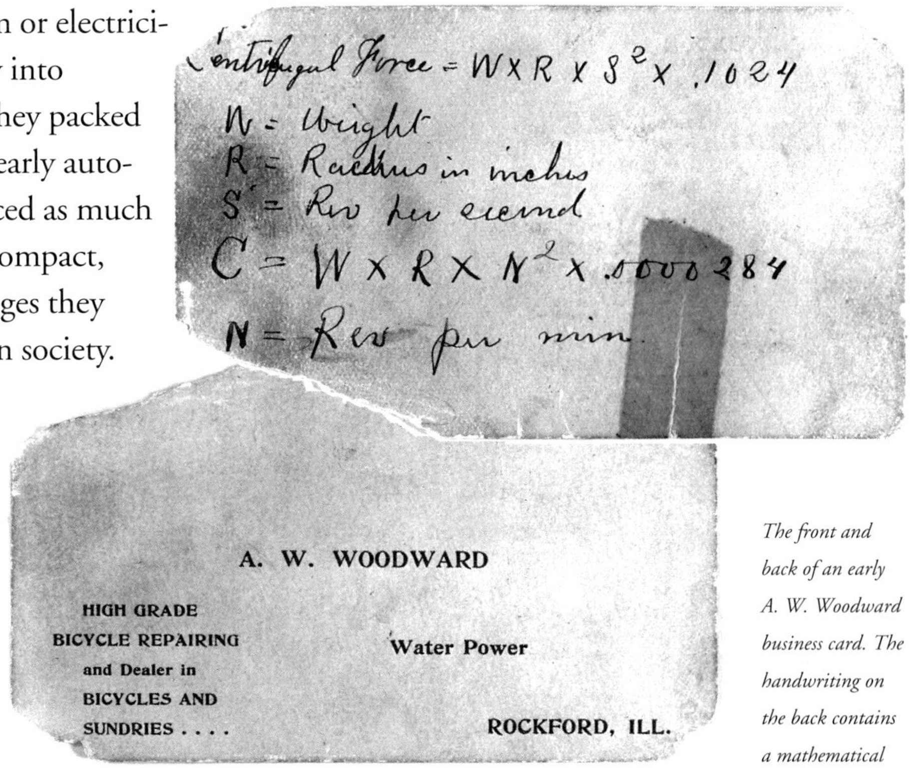

The front and back of an early A. W. Woodward business card. The handwriting on the back contains a mathematical formula for determining centrifugal force.

But in 1872, none of this could be foreseen. Water was still king, at the height of its power, even in the last years of its reign. Mechanized power was an industrial, not yet a social, force. But from then on, the United States would enter into a transition in power usage that would see its original reliance on wind and water, and the physical energies of men, women, and animals, converted into the open opportunities afforded by an increasing array of sophisticated machinery.[7]

In 1869, Amos Woodward was forty years old, and the United States was only seven years away from the centennial celebration of its Declaration of Independence. Like Amos, America had grown up, painfully at times, but was now in the full flower of its maturity. That year, Ulysses S. Grant was inaugurated as the eighteenth President. The Civil War, though ended five years earlier, had left a legacy of confusion and hatred. Blacks had been freed from slavery, which did not sit well with Southern radical groups like the newly formed Ku Klux Klan, but the South's real anger was now directed toward the policy of Reconstruction, which was often interpreted by some unscrupulous Northerners as an excuse for exploitation and personal enrichment. In an act of reconciliation, Grant allowed the remaining four Confederate states to rejoin the Union.

Abroad, revolutionary impulses, largely intellectual in character, were stirring. These offer a fitting contrast to the more controlled bursts of energy then taking place in America, exemplified by Amos's decision to place his economic future in his own hands. The latter half of the nineteenth century saw the birth of a widespread movement to apply the methods of the natural sciences, especially physics and mathematics, to the study of human beings and social and political groups. Out of this movement came the social and political sciences. Amos, however, belonged to an older, and distinctly American school of thought which the French social philosopher Alexis de Tocqueville, in his 1835 book *Democracy in America*, had called "practical philosophy." Tocqueville saw in this philosophy the American's habit of bending his mind and energies toward the solution of practical, everyday problems. The typical American was too busy populating an uncharted wilderness, cultivating fields, and building cities and towns to take interest in the elusive, abstract issues that occupied European academic life. This was why, he said, Americans who had not gone to college still rose to important positions in government and business, even if they were embarrassingly naive in the higher intellectual pursuits of abstract philosophy. Americans held fast to the belief that the individual man, by rights granted him by God and nature, was responsible for the conduct of his own life and the success or failure of his endeavors.[8] Thus, when a man like Amos started his own business, leaving a secure job behind him, he was simply following the basic tenets of the American practical philosophy.

Life was never so serious, though, that it required attention to the higher, dryer abstractions of philosophy. For Rockford had a baseball team, and a good one, led by pitcher Albert G. Spalding, a future major leaguer who later founded the sporting goods company that bore his name. In 1869, he won twenty-four games and lost only four, all to the barnstorming Cincinnati Red Stockings, the most powerful team in baseball. Once, in fact, the Forest City baseball team (as it was called) almost pulled off the impossible. In a game played in Cincinnati, Forest City led 14-12 in the ninth inning, but the Red Stockings scored three runs in the bottom of the ninth to win 15-14. Forest City was not taken seriously, a fact which the *Cincinnati Gazette,* in its headline for July 26, duly noted: "A Sucker Club Comes Within One of Dealing the Red Stockings a Socker."[9] This was the closest any team would come that year to defeating the Red Stockings. They were undefeated in over fifty-eight games, and their tour across America that year popularized baseball throughout the country.

In 1873, Amos continued to make governors, but he actually ran a general machine shop that made a number of products, including pump cylinders and Barnes jigsaws.[10] This was a perilously small business, certainly undercapitalized, just getting underway in difficult times. The national economy had remained strong since the

Civil War, except for a mild recession in the summer of 1869. Then, following the Panic of 1873, a major economic downturn occurred, caused by a crisis in banking and finance. Liberalized credit practices had been promoting most of the business expansion of the early 1870s. When this was combined with the tremendous railroad expansion then going on and the controversy over the Credit Mobilier scandal, which blackened the name of railroad promoters and several congressmen alike, the stage was set for a business contraction. It began in September 1873, when Jay Cooke and Company, overly-involved in the promotion and finance of the Northern Pacific Railroad, went under. Cooke's failure was a symptom of the general over-expansion in industry and agriculture then going on. By the end of the year, over 5,000 businesses would fail nationwide, and the economy would be depressed until 1879.

Despite the severity of the depression, it seemed to have little bad effect on Rockford. Its population had increased from about 7,000 in 1860 to over 11,000 in 1870, and a number of new businesses besides Amos's started in the early 1870s. The hosiery industry began in 1870 when the Burson and Nelson Company knitted its first sock automatically. W. F. and John Barnes Company incorporated in 1872, and the People's Savings Bank was established in 1873. The Rockford Watch Company and the Forest City Furniture Company had their origins in the summer of 1874.[11] In 1871, Andrew C. Johnson and J. P. Anderson began Rockford's furniture industry in their planing mill in the Water Power. In 1873, they were joined by L. D. Upson, and together they built a new furniture factory at the corner of Race and Mill streets. Their entrepreneurship was followed by many other, predominantly Swedish, residents, so that within twenty years Rockford became the second largest furniture manufacturing city in the United States, behind Grand Rapids, Michigan.[12]

By 1874, Amos was deeply preoccupied with the pressure of being the sole proprietor of his own business. The outlook was bleak: he needed more capital to run the business, and he needed a partner to help him. During 1874 he managed to acquire both.

On July 9, 1874, Amos remortgaged his home through the newly established People's Savings Bank. He had paid off his original mortgage in 1865, and the property had remained free from encumbrance since then. The new mortgage was for $2,000 in principal for three years at 10 percent per year payable semiannually.[13] This was quite a burden to take on, but it reflects his confidence in the potential of his business.

Also in 1874, William G. Ross, whom Amos had known since their days together at N. C. Thompson, joined Amos to form the short-lived partnership of Ross and Woodward. They proudly announced their new business in the 1874-1875 *Rockford City Directory* by taking out a full page advertisement that highlighted Amos's patented governor. But the boldest type noted that they were machinists and repairers. In addition to handling "all kinds of jobbing and repairing of machinery," they manufactured windmill, well, and cistern pumps.[14] Then, in 1876, the partnership dissolved, and Ross went back to his trade as a pattern maker. By this time, the governor part of the business was de-emphasized in favor of the iron pump business, in which "deep and shallow well force and suction pumps [were] a specialty." The company was still located east of the woolen mill in the Water Power District.[15]

Amos's relationship with his family in Maine continued much as it had in earlier years. He regularly sent his aged mother money and presents from 1868, when his father died, at least through 1874, and probably longer. She appreciated his kindness, but lamented in 1871 that "[I don't] think I shall ever see any more letters from my dear son Amos…[I] should like to hear about his patent Governor, hope he will come and tell us if he does not write any of us."[16] Despite this hopeful entreaty, the demands of his family, business and church were insistent, and he remained in Rockford.

The business remained small, Minnie (the last of four children) was born in 1874, and Amos's responsibilities with the First Baptist Church increased in the 1870s. On September 11, 1870, he was appointed to his first church offices: the Pulpit Committee and the Finance Committee. A month later, he was made a representative to a Council of Ministers and Delegates from sister Baptist churches to deal with the dissident Reverend Thomas Kerr, former pastor of the First Baptist Church. Kerr had left the church with forty-eight other parishioners to form the Church of the Christian Union, and this had caused a major split among the congregation. Eventually, the meeting that Amos attended "withdrew the hand of fellowship" from Reverend Kerr because he was a nonconformist.[17]

In 1871, Amos was named Church Treasurer, a post he was re-elected to in 1872. In 1873, he was named a delegate to attend the Baptist Association meeting in Stillman Valley, and he was elected to that association's Financial Committee. On June 6, 1874, he was appointed to another committee that was charged with making arrangements for the Rock River Baptist Association meeting, which was held at the First Baptist Church in mid-June.[18]

By the mid-1870s (coinciding with the short-lived 1874-1875 Ross and Woodward partnership) there was an eighteen-month period where Amos held no church office at all. But in 1876, he was named to the Music Committee—a favorite committee of his because of his long-standing love of singing and music. He continued in this committee throughout 1876 and 1877, but held no church office in 1878 or 1879. These were years of some financial distress for the church, and the minister's pay was in arrears. Amos personally was doing relatively well financially, because he paid off the three-year

$2,000 mortgage on January 9, 1878—six months late but fully paid.[19]

By 1880 both Amos and Mate Woodward held church offices. He was again on the Music Committee, and she was on a committee that collected and distributed funds for needy church members. They both remained active for the next few years. Eventually, Amos's hard work and concern for his church was justly rewarded when he was named Deacon in 1884. He would be re-elected twenty-nine times to this position before retiring in 1914 as Honorary Deacon for Life.[20]

Amos's business became a family operation in 1877 when Elmer dropped out of school at age fourteen to become a machinist's apprentice at his father's shop. He had to stand on a soap box in order to reach the lathe he was learning to operate. Later that year, evidence of his inventive mind first surfaced when he rigged his lathe with gadgets that allowed him to sit and watch the machine operate without constant supervision. The shop foreman thought otherwise and was convinced that Elmer "was wasting his time thinking up ways to get out of work."[21] Instead of apparent laziness, however, Elmer's innovations showed practical genius. The value of this ability to the company would be proven when Elmer became the chief inventor and guiding force for the company as it expanded and prospered.

Two years after Elmer began his apprenticeship, the family role in Amos's business expanded again when Herbert M. Woodward, the son of Amos's brother Franklin, joined the firm to learn the machinist's trade. He was twenty-one when he came to Rockford from Winthrop, Maine, and he lived with Amos and his family on Peach Street.[22]

Thus, there were three Woodwards in a company that had no more than six hands in the late 1870s and early 1880s. Moreover, while water wheel governors were a major product, they were still secondary to the general machine shop work and pump manufacturing. Not until 1882-1883 would water wheel governors be mentioned in a *Rockford City Directory* as the primary product.[23]

In 1877, *The History of Winnebago County* was published, and it dispensed with A. W. Woodward in two-and-one-third lines, noting that he made "pumps for the wholesale trade," and that a water wheel governor was "also one of his leading articles." Meanwhile, the three companies Amos had once worked for were given more coverage and general approval or praise. Emerson, Talcott and Company employed 100 to 200 men; F. H. Manny 75 men, and N. C. Thompson 175. All seemed to be prospering; Amos was holding his own.[24]

This situation remained the same going into the early 1880s. The Panic of 1873 had run its course, and the economy was beginning to expand. Farm production increased, manufacturing expanded, and railroad construction took off again. By 1881 U.S. exports were at a record high, and consumer prices continued the long-term decline that had begun in 1865 and would last almost until 1900.[25] If the average person had a job that paid steady wages, his financial security would improve a little each year. Then another recession gripped the economy from 1883 into 1885. Businesses contracted or failed and jobs were lost. The unemployed were not totally forgotten. A strong commitment to family, close-knit neighborhoods, and churches and private organizations provided relief for those who had lost their jobs.

One man who lost his job, and his company, was N. C. Thompson himself, Amos's last employer. Thompson's company had become dependent on the sales of reapers for its success. However, when two-wheeled reapers were invented, they took over the market, making the Manny reapers obsolete. To make up for this, Thompson then manufactured plows and cultivators, but the plow business languished. Thousands were made and sent out on commission, but many were returned and scrapped. Thompson overextended himself personally and overextended his bank as well. Finally, in September 1874, the bank suspended trade, and Thompson became the "talk of the town." Shortly thereafter he retired from his manufacturing business.[26]

Thompson fancied himself a poet, and in 1887 wrote a poem titled "Rockford" for the preface of a local history book published that year. Critics might scorn his ability, but the third verse of his poem was right from the heart:

Invention, Trade and Industry;
Have wafted to the nation's bound
Thy honored name, which stranger lips,
With praises true, have justly crowned.
The toiler and the rich, thou hast,
In peaceful stations, side by side
For wealth's occasions, equal shared,
The hearts of men, have satisfied.[27]

These lines do capture the spirit of the times. By the 1880s, Rockford was described as the "Lowell [Massachusetts] of the West" because of its extensive and diverse manufacturing facilities, and because it was the second leading manufacturing city in Illinois. Rockford people were proud of this comparison, which the same local history justified as "a title not sought through bigoted conceit, but justly earned by thrift and enterprise."[28]

While Rockford grew, Amos's company remained tiny in comparison to the city's larger manufacturing companies. At the end of 1883, the company employed six workers, on an annual payroll of $1,500 and a gross annual product of $5,000. Capital available for the company's use was $4,000, much of it raised in the 1870s and 1880s through Amos's continual re-mortgaging of his house and land on Peach Street. He did this at least four times between 1878 and 1889.[29] Nevertheless, Amos's business remained undercapitalized and weak throughout the 1880s. His main product, the water wheel governor, was a good one, but the company was still classified as a general machine shop that manufactured pumps and water wheel governors, and compared to the rest of Rockford, it was not doing well in either. Between 1880 and 1890, Rockford's population had grown from 13,000 to 23,500. One promoting publication claimed in 1883 that the capital invested in manufacturing was almost $5 million, that some 5,000 skilled workmen were employed, and that the gross product they turned out was over $10 million in value. These figures, for the time, are impressive. As one historian noted, Rockford "is a town of importance, a town of promise, a town to attract continual fresh capital and more and more people..."[30] But, unless circumstances were changed at A. W. Woodward, the time might soon come when Amos's small company would have nothing to contribute to the local economic figures.

Rockford quickly shook off the 1883-1885 recession, and prosperity abounded throughout the remainder of the decade. Much was done through innovation and entrepreneurship to modernize and improve the city. It took four years for Alexander Graham Bell's new invention, the telephone, to reach Rockford. In 1880, seventy-four were installed under the auspices of the Western Union Telegraph Company. This was clearly the most exciting event of the year. Despite sporadic service due to storms and technical problems, to have a phone was to join the modern world, even if you had to twist a crank to ring up central and then hope that the manual connection would go through. Although there were only 300 subscribers by 1888, the system grew rapidly in the 1890s.[31]

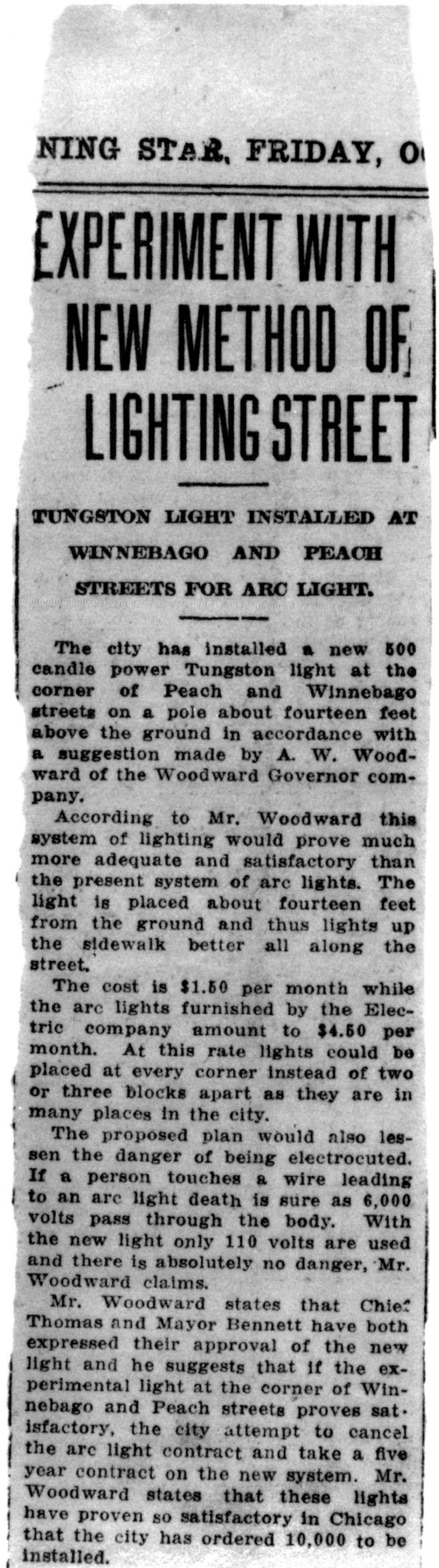

NING STAR, FRIDAY, O

EXPERIMENT WITH NEW METHOD OF LIGHTING STREET

TUNGSTON LIGHT INSTALLED AT WINNEBAGO AND PEACH STREETS FOR ARC LIGHT.

The city has installed a new 500 candle power Tungston light at the corner of Peach and Winnebago streets on a pole about fourteen feet above the ground in accordance with a suggestion made by A. W. Woodward of the Woodward Governor company.

According to Mr. Woodward this system of lighting would prove much more adequate and satisfactory than the present system of arc lights. The light is placed about fourteen feet from the ground and thus lights up the sidewalk better all along the street.

The cost is $1.50 per month while the arc lights furnished by the Electric company amount to $4.50 per month. At this rate lights could be placed at every corner instead of two or three blocks apart as they are in many places in the city.

The proposed plan would also lessen the danger of being electrocuted. If a person touches a wire leading to an arc light death is sure as 6,000 volts pass through the body. With the new light only 110 volts are used and there is absolutely no danger, Mr. Woodward claims.

Mr. Woodward states that Chief Thomas and Mayor Bennett have both expressed their approval of the new light and he suggests that if the experimental light at the corner of Winnebago and Peach streets proves satisfactory, the city attempt to cancel the arc light contract and take a five year contract on the new system. Mr. Woodward states that these lights have proven so satisfactory in Chicago that the city has ordered 10,000 to be installed.

Newspaper article (circa 1913) citing Amos's role in improved public lighting for sidewalks.

In another effort at modernization, the Rockford Street Railway Company was organized in late 1880 and began service a year later. This rudimentary effort at mass transportation used horses, since electricity was not yet available. The end result was not particularly satisfactory. Service was slow and the horses were sometimes unreliable and always smelly. By the end of the decade, the Rockford Street Railway Company had gone bankrupt, to be replaced in 1890 by a company that offered the modern convenience and speed of all-electric streetcars.[32]

Electricity had come a long way in the decade since Edison had perfected the incandescent light bulb in 1879. Shortly thereafter, he sought to build a central power generating station that could send power to illuminate people's homes. He succeeded in 1882 setting up the Pearl Street Station in New York City, shrewdly located to include much of the Wall Street financial district since he would be in need of large amounts of capital. His success there led to the organization of electric companies and generating plants throughout urban America. In 1883, Rockford built its own when the Forest City Electric Company was established to produce the electricity that would take the place of kerosene lamps and gaslights. In a burst of enthusiasm, one elated Rockford merchant described electric power as "that marvelous electric fluid."[33]

These modern improvements would, in the long run, transform Rockford as would other local developments that made the city a better place to live. Some 1880s highlights included the establishment of a hospital in 1883, the reorganization of the public school system in 1884, the founding of Central High School in 1885, the beginnings of a sewer system in 1886, and the first paved streets in 1889.[34]

It was also in the decade of the 1880s that Elmer Woodward finished his apprenticeship, continued working for his father as a machinist, fell in love, married, was named a foreman at the company, and began his own personal study and experiments with governors which would prove so fruitful to the success of the company in the future. Although he lacked a formal education, he was bright, well read, and studious. He became particularly adept at reading technical books, and in the process became what people today would describe as a "workaholic." Virtually every night, he studied or tinkered with inventive ideas on his drawing board at home. Much of the time he looked for inspiration by whistling "Here We Go Round The Mulberry Bush" under his breath, usually off key.[35]

In 1886, when he was twenty-four, Elmer fell in love with Mary (May) Fuller Dunham, an attractive nineteen-year-old photographer's assistant from Freeport, Illinois. She was an orphan who had been brought up by her grandmother, the formidable Mrs. Mary Hess. May had met Elmer through his sister Cora, and they carried on a tentative and affectionate correspondence while seeing each other whenever they could. By December 1886, Elmer surprised himself by asking for, and getting, Mrs. Hess's blessing to marry May. They planned their wedding for February 10, 1887, in Freeport, with the Woodward family pastor, Mr. William A. Stanton of the First Baptist Church in Rockford, officiating.[36]

As his wedding approached, Elmer's wry sense of humor emerged. He wrote his fiancee on Sunday, February 6, 1887 that:

> *We got an order for some more machines yesterday and business is just booming. I don't know whether I will be able to keep my engagement with you Thursday or not. If I can't I will send a proxy. However I think I will be there if I don't forget it or miss the train.*[37]

He was kidding, of course. But then, on February 8, Rockford was deluged by a heavy rainstorm that raised Kent and Keith creeks to their highest levels in forty years. The flood damaged the J. P. Manny and N. C. Thompson properties in the Water Power District. It also washed out the railroad tracks in that part of town. No railroad, no Freeport, no wedding. Fortunately for Elmer, these tracks did not serve the Freeport line. On February 10, he arrived safely, and on time, in Freeport, and the marriage went off as planned.[38]

The newlyweds apparently moved in with Amos's family at first, but within two years they rented a home at 511 Peach Street, one block from the family home at 603 Peach Street. Amos also promoted Elmer to foreman in 1887. This increase in pay would enable him to purchase needed groceries, like butter at 15¢ a pound and sugar at sixteen pounds for $1. Savon soap was twelve bars for 25¢ and gentlemen's shoes cost $2.75. If he wanted to treat May to a night on the town, he could buy a suit for $25 and take her to the Opera House, where tickets cost from 15¢ to 50¢ per seat. If they kept up with the news, they would learn that President Cleveland, who had also recently married, signed the Interstate Commerce Act. This act made railroad rates fairer, and would particularly benefit small companies like A. W. Woodward, which shipped goods by rail but could never qualify for rebates or drawbacks because of their small size. Amos's wife Mate probably was made more aware of her English origins that year, because the British were celebrating Queen Victoria's Fifty-year Jubilee.[39]

Mary ("May") Woodward, Elmer's wife, with baby daughter Dorothy who was born December 22, 1896.

Elmer and May's marriage turned out to be a very happy one, and they eventually had three children—Walter, born in 1889, Harold in 1892, and Dorothy in 1896. However, Elmer was often a frustrated young man in the first year of his marriage, when May was frequently absent to take care of the ailing, but still formidable, Mrs. Hess in Freeport. Elmer humored her, cajoled her, sometimes pricked her conscience into returning to him as soon as possible. Only two weeks after their marriage she was in Freeport, and he wrote her:

Don't you think you had better come home this week... it don't look just right to desert your husband after two weeks of war. God thinks you should also. See I Cor. 7-10. You can come home and stay a few days and then go back again if you wish.
Your affectionate hubby
Elmer[40]

She still hadn't returned by March 8, and he wrote a more testy letter that was softened by his closing:

Did you ever tell the truth except by accident. Everybody wishes to know when you will be home. I told them Wed. so don't make me out a l__r now... Come home as soon as convenient not that I wish to see you but it looks bad.
your Loving Husband
Elmer[41]

This time it worked. She came home, but by summer was back in Freeport. Now May took her turn at persuasion.

My Dear Elmer,
I do wish you would come up Saturday night. I want to see you awful bad, so bad. Last night I commenced to kiss Cora [Elmer's sister] in my sleep and she thought I was getting rather affectionate and asked me what was the matter and I said I thought you were Elmer.

Elmer replied by letter that same day:

Don't kiss Cora any more but save all your kisses for me when I come home.[42]

Elmer was a patient man. Two days later, using excerpts from a conversation he overheard between his next door neighbors, Mr. and Mrs. Hill, he wrote to prod May into returning. The Hill's conversation had made him lonely, and then he stated:

I wish my wife thought as much of me as she does of him. Mine goes away for a month and thinks nothing of it. Say are you not about ready to come home anyway[?][43]

Theirs was a lifelong, loving, and happy marriage only occasionally affected by Elmer's preoccupation with his job. One such occasion took place in February 1888, when May was again visiting her grandmother. Elmer wrote her a very matter-of-fact letter noting the weather and family events, and the equally neutral fact that he "was sorry for [his] sake" that she would not be home for a few more days but that they both "ought to be glad if [May's] visit [gave] grandma any pleasure." It was a normal chatty letter, until he signed it "Yours Truly, A. W. Woodward." He added in a postscript:

Darn it, I forgot for a minute
I was not writing a business letter.
Ever your affectionate Husband Elmer.[44]

In 1886, when Amos's small company was well into its second decade, the United States was about to enter its second century under the Constitution. It was a big country of forty-four states (Arizona, New Mexico, Utah, and Oklahoma were still territories) stretched out from coast to coast. The census of 1880 had put the population at over fifty million, occupying a total land area of more than three million square miles. Tourists were flocking to the world's first national park, tucked away on an isolated mountain plateau along the Yellowstone River in northwest Wyoming. Thanks to the telegraph, news came from all over the country. In Chicago, the first serious anarchist agitation in America erupted into riot. During a rally in Haymarket Square, a bomb exploded among the police, who then fired upon the demonstrators. Arrests were made, and the leaders of the rally, including August Spies and Albert Parsons, were convicted of conspiracy to provoke murder, and executed. Farther away, in Arizona, Apache war chief Geronimo was captured and returned to the reservation. And fittingly in this year, 1886, the close of America's first century, the Statue of Liberty was dedicated.

The biggest news in Rockford that year was the collapse of Amos's old employer, the N. C. Thompson Company. Its bank failure of two years earlier had already ruined its credit. Then disaster struck when a fire destroyed the Johnston Oat Meal Company and several nearby buildings, including Thompson's warehouse. In late December, the company folded for good. The *Rockford Daily Gazette* noted:

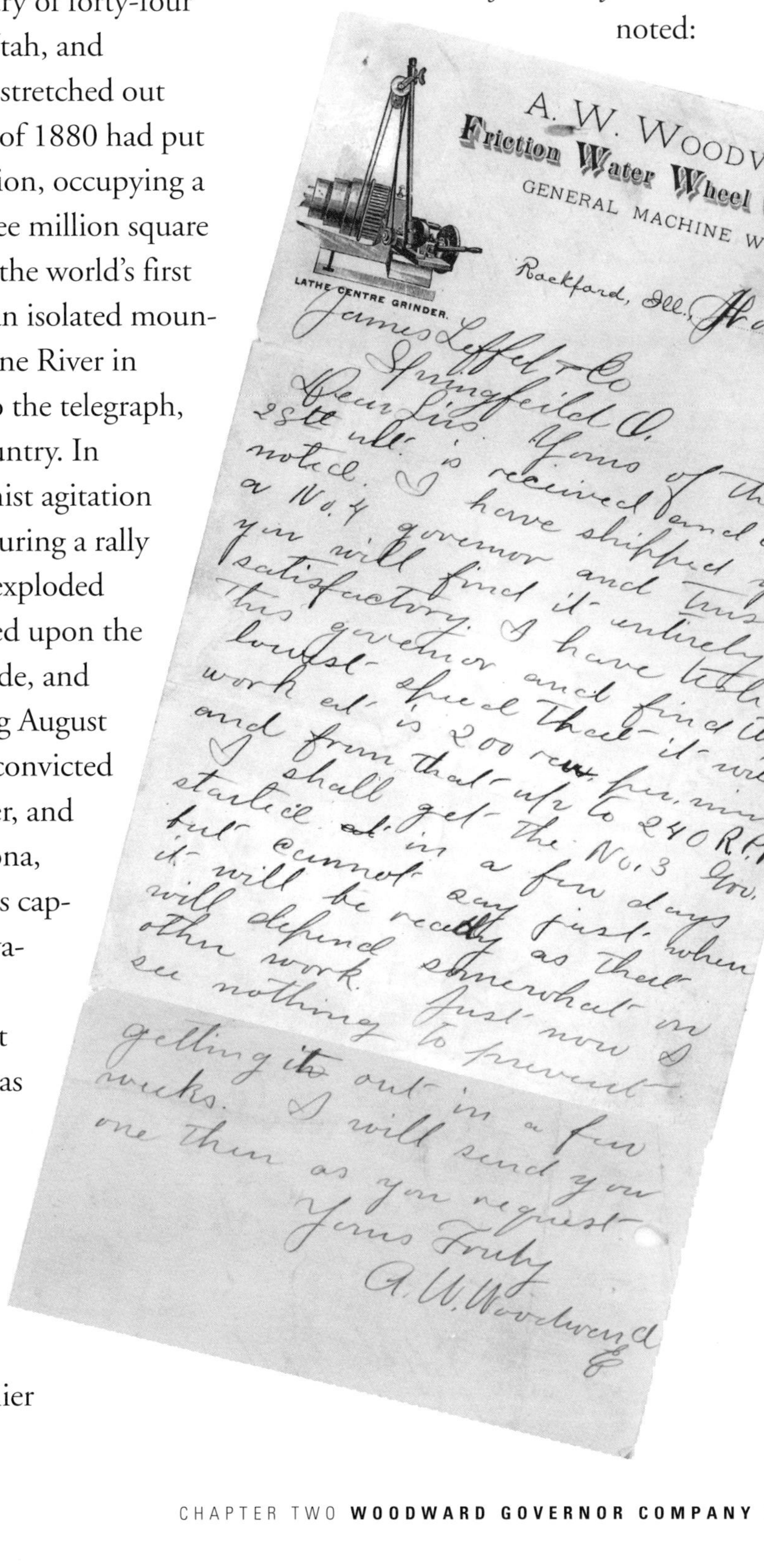

A. W. WOODWARD,
Friction Water Wheel Governors,
GENERAL MACHINE WORK.
LATHE CENTRE GRINDER.

Rockford, Ill., Nov 2 1887

James Leffel & Co
Springfeild O.

Dear Sirs: Yours of the 28th ult. is received and contents noted. I have shipped you a No. 4 governor and trust you will find it entirely satisfactory. I have tested this governor and find the lowest speed that it will work at is 200 rev. per min. and from that up to 240 R.P.M. I shall get the No. 3 Gov. started in a few days but cannot say just when it will be ready as that will depend somewhat on other work. Just now I see nothing to prevent getting it out in a few weeks. I will send you one then as you request

Yours Truly
A. W. Woodward
E

The earliest business letter on Woodward stationery, dated November 2, 1887. Though the letter was signed by Amos, the handwriting is by Elmer, who had joined the company in 1877 at age 14.

"It has been known for some time that the company was not doing any business."[45] Amos, it turned out, had made the right choice to leave when he did.

Despite such setbacks, by the end of the decade Rockford was booming. The *Daily Gazette* boasted that over 4,500 men were employed in the city's principal factories. And this number did not include storekeepers, clerks, and artisans. A. W. Woodward was one of nine firms listed under foundries and machine shops in the paper's survey. The largest was W. F. and John Barnes, with 148 employees, and the smallest was A. W. Woodward with four employees—down two from the last survey earlier in the decade. One of those two who had left was Amos's nephew, Herbert, who had gone to work as a machinist for Savage and Love in 1889.[46]

Despite the general prosperity, Amos's machine shop and governor business was not thriving. The governor had not been improved much since the original patent in 1870, and its clientele was limited. By 1889 only about 500 governors had been sold since the first one was made in 1869. The average of twenty-five governors sold per year was not enough to run a successful business, unless other products were sold as well.[47]

It must have been frustrating for Amos and Elmer to struggle along so steadily while Rockford grew so rapidly and other industrialists did so much better. The *Rockford Daily Register's* headline for December 31, 1889, was "Our Great Growth." The accompanying story described Rockford not only as "The Lowell of the West," but as the coming "Manchester [England] of America." Dozens of industries were noted in the article, but no mention was made of Amos's company. A few weeks later the paper noted the big boom in real estate values in Rockford.[48] Lots of money was being made, but very little of it was finding its way to Amos or Elmer Woodward.

Something would have to be done to improve the water wheel governor and thereby enhance the company's chances for expansion and financial success. Fortunately Elmer's inventive mind had long been at work with plans to improve their products. These plans would become reality in 1890, when Amos and Elmer jointly received a patent for a faster governor that led, at least initially, to a rapid rise in their business and a general rejuvenation of the company during the early 1890s.

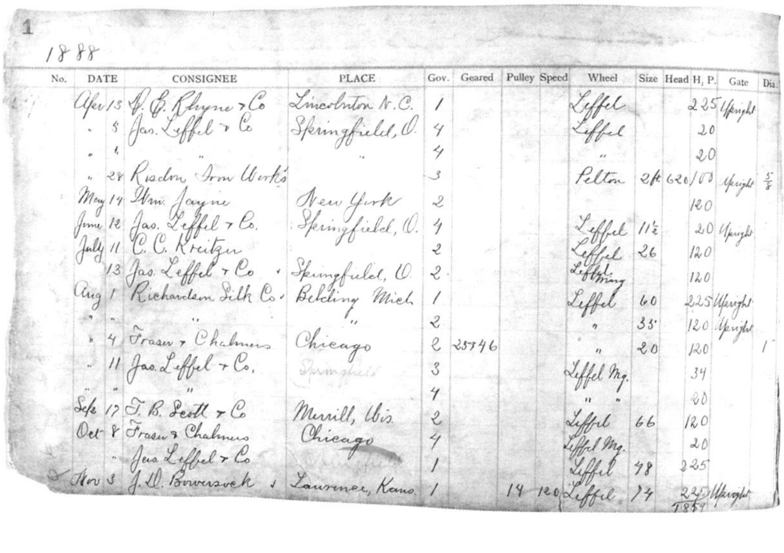

1

1888

No.	DATE	CONSIGNEE	PLACE	Gov.	Geared	Pulley	Speed	Wheel	Size	Head	H. P.	Gate	Dia.
	Apr 13	J. E. Rhyne & Co	Lincolnton N.C.	1				Leffel			225	Upright	
	" 5	Jas. Leffel & Co	Springfield, O.	4				Leffel			20		
	" 6	"	"	4				"			20		
	" 28	Risdon Iron Works		3				Pelton	2 ft.	620	100	Upright	5/8
	May 14	Wm. Jayne	New York	2							120		
	June 12	Jas. Leffel & Co.	Springfield, O.	4				Leffel	11½		20	Upright	
	July 11	C. C. Kreitzer		2				Leffel	26		120		
	13	Jas. Leffel & Co	Springfield, O.	2				Leffel Mfg			120		
	Aug 1	Richardson Silk Co.	Belding Mich	1				Leffel	60		225	Upright	
	" "	"	"	2				"	35		120	Upright	
	" 4	Fraser & Chalmers	Chicago	2	25746			"	20		120		1"
	" 11	Jas. Leffel & Co.	Springfield	3				Leffel Mg.			34		
	" "	"		4				" "			20		
	Sept 17	T. B. Scott & Co	Merrill, Wis.	2				Leffel	66		120		
	Oct 8	Fraser & Chalmers	Chicago	4				Leffel Mg.			20		
	"	Jas Leffel & Co		1				Leffel	48		225		
	Nov 3	J. D. Bowersock	Lawrence, Kans.	1		14	120	Leffel	74		225	Upright	
											1854		

The first page of the earliest extant Woodward "serial book," showing sales of governors for 1888.

The Springboard to Modest Success

1890-1902

The decade of the 1890s was a transitional one for America and for the A. W. Woodward Company. It started out promisingly enough, until the Panic of 1893 led to a general depression that was to last over four years and would devastate much of the country's business and industry. A precipitous drop in governor sales forced Amos and Elmer to try bicycle sales and repairs to tide the company over. But they quickly abandoned this line of business once the economy improved, although the company would not really thrive until late in the decade when Elmer perfected his new compensating governor. This invention would put the A. W. Woodward Company in the forefront of the governor industry, rapidly increase its reputation and sales, and ensure its success through the first decade of the twentieth century.

Due to Amos's experience and Elmer's ingenuity, their joint patent of 1890 was followed by four other patented governor improvements between 1890 and 1901. This first patent was immediately successful, as governor sales expanded rapidly during the first three years of the 1890s. Elmer was enthusiastic about the company's prospects, even if it meant giving up one of its non-governor sidelines. Writing to May, he noted that Will Allyn, one of his friends, was home again and that he was "going to make a lot of carpet machines for his father so that lets me out of a job." This loss of an auxiliary business might have posed a problem in earlier years, but, Elmer continued, "if the governors sell the way they have for the past few months we won't have any time to make [carpet machines] anyway." May's reply was equally enthusiastic: "I am glad you are having such good luck with your governors. Mr. Emmert said that the Woodward governor was the best governor in the U.S."[1]

During this time Elmer was working a six-day week that often continued into Saturday evening. The extra hours and hard work began to pay off as governor sales jumped from thirty-six in 1889 to fifty-nine in 1890, sixty-six in 1891, and seventy-six in 1892. Still, the company remained small. The capital invested was about $7,000, and annual sales of $8,000 in 1891 allowed for only three employees and a company payroll of about $2,000.[2] Although this was worthy of mention by Rockford boosters, it paled in comparison to the multitude of other industries that had made the Forest City the "Lowell [Massachusetts] of the West."

But the city's fortunes were favorable to A. W. Woodward's future success. To Rockford promoters, the city was "a queen amid queens," the "industrial city of the west, a wonderful manufacturing municipality," and the home of "thirty thousand progressive people." Poetry captured the spirit that plain prose and statistics could not:

The din of commerce greets
the listener's ear,
Hammer and anvil sound
an echo clear,
The sturdy tread of
honest labor's foot
Tells where the oak of progress
finds a root.[3]

Rockford's status as an important industrial city was mirrored in its growing social consciousness. The P. L. Philippi Company published the city's first Blue Book, the "Fashionable Private Address Directory and Ladies' Visiting List." It contained the names of over 700 prominent householders, including Amos and Elmer Woodward. Although listed among the socially prominent, it does not appear that either Mate or May Woodward set aside the usual weekly receiving day for home visits by the local "blue bloods."[4] This is not surprising, since both women were private, family-oriented people, not society matrons.

Other examples of progress abounded in Rockford in 1890 and 1891. Business continued to improve, and new industries located in the city. Shoppers could buy Eagle milk for 18¢ a can, peanuts for 10¢ a pound, handmade chocolates for 18¢ a pound, graham crackers for 11¢ a pound, and a one-pound can of lobster for 28¢.[5] Sewers, described as "pretty expensive luxuries," were installed, and the street railway system made its mules obsolete when it electrified.

Nationally, another use of electricity attracted public attention when New York State installed and used electric chairs in its prison system. After the second execution was carried out at Sing Sing Prison in July 1891, the *Rockford Daily Register Gazette* editorialized that,

It is a slight disappointment to public curiosity that the penalty is made strictly private, but the educational influence which it conveys is certainly a healthy one in contrast to the spectacle sometimes made at an execution…

The spectacle of a crowd witnessing a man's death was replaced by a more discreet form of capital punishment that used "man-made lightning" withdrawn from public view.[6]

For the Woodwards, the early 1890s were prosperous and encouraging years. Sales not only expanded, they became more geographically

CATALOGUE AND PRICE LIST

WOODWARD'S....

NEW FRICTION

Water Wheel Governor.

MANUFACTURED BY

A. W. WOODWARD,

ROCKFORD, ILLINOIS, U. S. A.

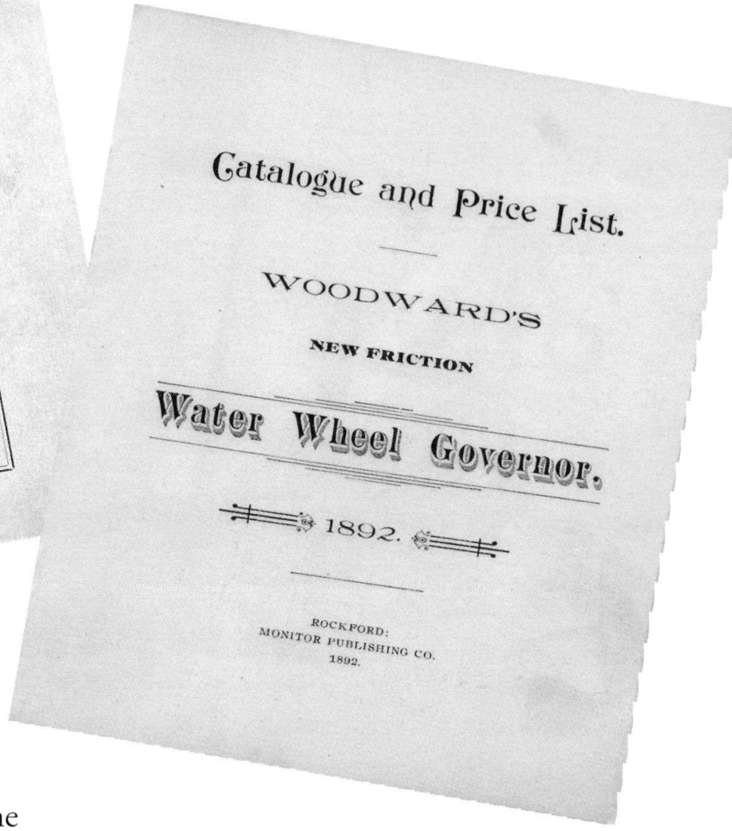

Catalogue and Price List.

WOODWARD'S

NEW FRICTION

Water Wheel Governor.

1892.

ROCKFORD:
MONITOR PUBLISHING CO.
1892.

Woodward's new friction water wheel governor, catalog, and price list, 1892.

diverse. In 1888, sales had been confined to the Midwest, and in 1889 sales were mostly to the western states. By 1890, when sales went from thirty-six to fifty-nine governors, purchasers were located all over the United States. This trend continued in 1891 and 1892; governor sales increased from sixty-six to seventy-six governors by 1892. The majority of the governors were applied to Leffel water wheels and turbines, although a few were for Victor, Davis, Flermiken, or Pelton water wheels.[7]

The A. W. Woodward Company's relationship with The James Leffel & Company, which was to last for decades, was an advantage to both companies. Leffel had begun manufacturing turbines in 1862 in Springfield, Ohio, and ultimately produced a novel and superior product—the double turbine. It was a model of efficiency and sold well into the 1920s, even though James Leffel himself appeared to be at a loss to explain why the turbine worked so well.[8] Elmer would probably have known, but he was much more concerned with improvements on Woodward governors.

In 1892, A. W. Woodward published a catalog and price list for the company's new "friction" water wheel governor, which replaced the 1890 patented governor the company was discontinuing. The catalog carefully explained governing principles and noted the importance of a "quick-acting" governor to keep speed nearly uniform when the gates of the wheel were opened or closed. The company stated that its governor was "powerful," "simple, strong and durable," and "reliable and time-tried." It emphasized that the new governor would operate the gate in one-third the time usually needed. It could do this because its motion was constant and uniform, not choppy and intermittent. The purchaser was encouraged to have "no doubts" about the new machines because Woodward's past successes and extended experience ensured "perfect results in every case." That was quite an implied guarantee, but the company from its origin always stood by its products and guaranteed that they would work.[9]

The new governors were made in four sizes that enabled their adaptation to any sized water wheel. No. 4, the smallest, was designed to regulate wheels fifteen inches and under in diameter. It was twenty inches in height, eighteen inches in length, and eight inches in width. It weighed sixty pounds and sold for $50. The No. 1 governor, the largest, was for wheels fifty inches and over. It was thirty-six inches by thirty-four inches by eighteen inches, weighed 285 pounds, and sold for $125.[10]

The catalog concluded with a three-page list of testimonials from satisfied customers. Woodward's older governors had been good machines, but the new friction type was even better. It is clear from the customers' enthusiastic letters that the product's future was promising, and that Woodward and Leffel had developed a mutually beneficial relationship because so many letters referred to Woodward governors on various-sized Leffel wheels.

The year 1893 was a pivotal one, with the company expanding into a larger building. Amos, who was a cautious man, was involved in a mental tug-of-war in early 1893, between his wish to relocate his expanding business, and his fears that expansion needs might be temporary. His tiny business might fail if the economy turned bad or if competitors came up with improved products. Certainly, on the basis of its increased sales, the company needed more space to build its governors. Also, the national economy seemed buoyant, and Rockford leaders were confident that the city's prosperity would continue. Rockford was in a building boom. Over 200 homes were under construction, the Nelson House Hotel was nearing completion, and many of the city's 187 factories were contemplating expansion.[11] The time seemed right for A. W. Woodward to make the move to larger quarters.

Consequently, Amos and Elmer rented the building formerly occupied by the N. C. Thompson's Reaper Works, a larger three-story building at 658-660 Race Street in the Water Power District. Before doing so, Amos did his best to ensure that, if the venture failed, his family would not be brought down with him. Therefore, he deeded his house at 603 Peach Street to his wife.[12]

After moving to Race Street, the company prospered in the first half of 1893, and for a time it appeared that the 1892 record of seventy-six governors sold would be easily exceeded. By October 28, the company had already sold sixty-three governors. But that was the last date it would sell a governor for more than three months.[13] The governor market had completely collapsed due to repercussions from the Panic of 1893.

The former Thompson Reaper Works (rear view), as it appeared in 1893, when Woodward moved into the building at 658-660 Race Street.

The country had been prosperous since the mid-1880s, and the economy had been expanding for seven years. Foreign trade was strong, and large amounts of European capital had been invested in a wide range of American businesses. Then, in early 1893, the New York Stock Exchange began a sharp downward trend. This was followed, beginning on May 5 and accelerating after June 27, by a genuine panic to dispose of stocks, bonds and other securities. The panic started when foreign investors withdrew their gold and sold off their American securities. When gold reserves dropped below the $100 million level, which many experts felt was the minimum necessary reserve to support the government, people lost confidence in the soundness of the dollar.[14]

The Panic of 1893 began a period of almost four years of the most severe depression the nation had faced up to that time. Recovery would not begin until 1897. In the meantime, millions of people suffered across the nation, and private charity was practically overwhelmed by the people's needs. During 1893 alone, 15,242 firms went bankrupt in the United States. In Rockford, twenty-six factories out of the total of 187 went into receivership in one day, and hundreds of jobs disappeared. Charitable societies opened soup kitchens that charged a penny for a portion of soup and bread. As late as 1897, Rockford's mayor commended The Ladies Union Aid Society for helping the unemployed help themselves through such means as "The Potato Patch Plan," which allowed the unemployed to convert empty lots into vegetable gardens.[15]

Despite the severity of the depression, the A. W. Woodward Company survived until renewed confidence in the economy, and its own inventions, led to an increase in business and ultimately the decision in 1902 to incorporate. In the meantime, the mid-1890s were rough years. Sales completely stopped for three months between late October 1893 and late January 1894. When they resumed, orders came slowly for the next eighteen months. Only eighteen governors were sold in 1894—the lowest number since 1888 (the earliest year for which sales records are available), and perhaps the lowest in the history of the company. Sales were down by more than 75 percent from the peak year of 1892. The situation only barely improved throughout the first seven months of 1895, when fifteen governors were sold. After August 1, sales picked up considerably when thirty-nine governors were sold in the last five months of the year. However, the upward trend did not continue, and sales declined to thirty-nine governors sold in 1896 and twenty-eight in 1897—the year the national economy began to return to prosperity.[16]

Caught in the middle of a depression, Amos and Elmer searched for a way to keep the company afloat. They thought they found it in 1894 when they demoted the governor business and emphasized, at least in the *Rockford City Directory*, the company's bicycle and machine shop business. They stressed that they did "job work and high-grade bicycle repairing." They also sold "bicycles and sundries." Only at the end of their advertisement did they mention that they manufactured water wheel governors.[17]

A newspaper advertisement for the A. W. Woodward bicycle repair business. The ad is dated from 1894, when the country was in the midst of a depression, and Woodward had to turn to bicycle repairs to keep the company afloat.

Bicycles, with the collapse of the governor market, were a smart choice, for bicycling had become a nationwide craze by the mid-1890s. It increased people's mobility and sociability and, with the advent of the safety bicycle with pneumatic tires, bicycles were considered safe and entirely proper for women as well as men. In fact, women were encouraged to think of bicycling as a way to get moderate, healthful exercise. A popular song, "Daisy Bell," with her "bicycle built for two," captured the nation's imagination and promoted the founding of bicycle clubs. A number were formed in Rockford, and both men and women participated in outings, and even in bicycle races held at the one-mile track at the Rockford Driving Park or on the smaller track at the Fairgrounds Park.[18] Bicycle racing was a nationwide, overwhelmingly male sport in which speeds increased rapidly as bikes were improved. In the pre-horseless carriage era, "scorchers" thrilled the hearts of young and old.

Amos and Elmer took advantage of this interest to ride out the depression, but their emphasis on bicycles lasted less than two years. Bicycles were only a stopgap measure, and would be quickly abandoned once the governor business recovered.

In the meantime, Amos and Elmer worked especially hard on increasing the speed and reliability of their governors through improvements that could be patented. They applied for a joint patent on October 5, 1896, and it was granted June 1, 1897. Shortly after, Elmer was ready with a further enhancement. The patent he applied for on July 3, 1897, was granted on August 2, 1898.[19] This patent added compensation to the governor, a major improvement that allowed faster gate movement without "racing" or overshooting the desired change in gate position. Accuracy of governor control also improved with this innovation.

These inventions were great improvements in the company's basic governor design over the earlier non-compensating models. Elmer improved his governor again with a patent granted in 1901.[20] The company thrived as a result of these developments. Never again would it stray from its original purpose and emphasis on building better speed control devices.

The end of the century was truly a new beginning for the company because these three patented improvements coincided with the return of a strong U.S. economy and an increased interest by the United States in the outside world. The depression had ended in 1897, thanks in part to foreign demand for American grain, and to gold discoveries in Alaska and Northern Canada that increased the gold supply. Moreover, the United States was taking on a more active role in international affairs. Following the Spanish-American war, fought to eliminate Spanish domination in the Caribbean and South Pacific, the United States acquired Cuba, Puerto Rico, Guam, and the Philippine Islands as protectorates. Also, Wake Island was acquired and the Hawaiian Islands were annexed in separate treaties.

These were heady years for many Americans who shouted "Remember the Maine" and cheered "Teddy" Roosevelt and his Rough Riders for their exploits at San Juan Hill in Cuba. Although President McKinley and his advisers cast a wary eye on Asia, others thought that the region, with its hundreds of millions of people, would serve as the new frontier for American trade and prosperity. The Open Door Notes, which promoted the equality of commercial opportunity in China, followed, but the trade increase was illusory and very limited.

In 1900, the Currency Act declared the gold dollar to be the standard unit of value and placed all forms of money issued by the government on a parity with gold. This antagonized the free silver Democrats, who had spent most of the previous decade promoting the free and unlimited coinage of silver at a 16:1 ratio with gold.

They, and the anti-imperialist bloc opposing international expansion, supported William Jennings Bryan for President and Adlai Stevenson for Vice President. However, McKinley with his slogan, the "Full Dinner Pail," and his Vice President, Theodore Roosevelt, carried the election, and the Republicans also retained their majorities in the House and Senate.

In Rockford, Amos and Elmer Woodward saw their company expand and prosper due to the favorable business climate and, above all, to Elmer's turn-of-the-century inventions. Governor sales had been flat or slightly down in 1897, 1898, and 1899, averaging only thirty-six per year. However, by that time governors had been sold in over thirty states from Maine to Washington and Texas to Minnesota. Most had been sold in the Midwest, but the east coast was well represented, and there were scattered sales throughout the west.[21]

By 1900, Elmer's new compensating governor began to pay off. It had reduced the time necessary to move the water wheel gate through its total stroke to about five seconds, compared to the five minutes for Amos's original 1870 governor and, two minutes for the non-compensating governors of the 1890s. According to company legend, during a demonstration of his son's new invention, Amos thought the test machine was running out of control and tried to stop it. He was quickly disabused of that notion and subsequently amazed by the result.

Sales more than doubled, from thirty-six in 1899 to seventy-five in 1900. Moreover, especially after the compensating governor's invention, the A. W. Woodward Company's domestic reputation had spread to Europe so that, in 1898, Woodward began to acquire regular overseas customers. One of these was the Alessandro Calzoni Company of Bologna, Italy, which bought its first machine from Woodward in February 1898, and bought two more before the year was over. In 1899, Calzoni accounted for eight of the thirty-six machines sold. Calzoni continued to increase in importance to Woodward through 1900, and purchased twenty-nine of eighty-eight sold in 1901. In fact, overseas sales began to make up a significant portion of each year's total sales beginning in 1899. By 1900, overseas sales were 30 percent of total sales, and this increased to over 40 percent in 1901 and 1902. Total sales exceeded 100 governors for the first time in 1902. Woodward governors continued to be sold throughout the United States, but by then there were regular, repeat customers in Italy, Finland, France, Sweden, and Canada. In fact some of Woodward's United States customers were reshipping Woodward governors to Russia, Peru, and Chile.[22]

Because of the increased sales and profits that began in 1900, the company had by necessity begun to expand. By 1905, there were seventeen employees who were able to manufacture and sell over 200 governors. Even before this expansion,

Amos and Elmer decided to incorporate the company. This was accomplished in 1902, and the A. W. Woodward Company became the Woodward Governor Company with Amos as president and Elmer as vice president. Three hundred shares of stock were issued, with Amos and Elmer each owning 125, and the secretary of the company, J. Charles Young, owning the other 50. After incorporation, the company would continue for decades to be small, but it would never revert back to the tiny dependent company of its first three decades. The early 1900s would bring continued expansion, higher profits, and a higher standard of living to the owners and members of the Woodward Governor Company.

The Woodward Governor Company's corporate charter, issued by the State of Illinois on June 7, 1902.

Incorporation to Construction

1902-1910

When the Woodward Governor Company incorporated in 1902, it was capitalized at $30,000. Each of the 300 shares of stock was worth $100 par value. For the next four decades the company would remain capitalized at $30,000, although its growth and net worth would soon dwarf its original capitalization. Even in the first decade of the twentieth century, tremendous strides within the company brought about the kind of prosperity that must have astounded Amos Woodward. He had worked hard for three decades and achieved his measure of success but, from the time the company incorporated in 1902, until his death in 1919 at age eighty-nine, his life would be made comfortable thanks to the regular dividends he received from his ownership of a large plurality of the company's stock.

Amos remained president of the Woodward Governor Company until his death, but he was progressively less active in the actual control of the company after 1902. Elmer was chiefly responsible for most of the company's successes during these years. He not only handled the business, but kept up with new developments to maintain the company's lead in the governor field. He also oversaw the company's growth during the first decade of the twentieth century. By 1910, the need for increased space forced the company to relocate. After four decades in rented quarters, the company built its own facility.

The general prosperity of the era almost certainly aided the prospects of the small company. Throughout the United States and in Rockford, people felt a sense of renewed confidence with

Mary (Mrs. Elmer) Woodward, with children Dorothy and Walter, in front of their house at 314 North Winnebago Street, Rockford, probably 1902.

Two of Rockford's most prominent citizens, Amos and Mary Woodward. The photo was taken behind Elmer's house at 817 Haskell Avenue, which was purchased in 1907.

the new century. America was becoming a world power. In 1899, the war with Spain was formally ended when the Senate ratified a peace treaty. Under the treaty's terms, Spain gave up control over Cuba (which became a U.S. protectorate until 1902), and surrendered Guam, Puerto Rico, and the Philippine Islands to the United States. Also in 1899, Secretary of State John Hay's Open Door Notes, addressed to other world powers, led the way for U.S. trade with China on a basis of equality with other nations.[1]

Theodore Roosevelt succeeded to the presidency after the assassination of William McKinley in 1901, and it was Roosevelt who oversaw America's rise to world prominence. Among his more daring initiatives was the Roosevelt Corollary to the Monroe Doctrine, which he presented before Congress in his 1905 State of the Union Address. The Corollary stated, in principle, that the United States would tolerate no outside interference in Latin America. It further asserted America's obligation to take part in Latin American affairs, if the governments there requested intervention, or appeared in need of it.

At home, Roosevelt's rhetoric was often more restrained, and his successes were more enduring. He entered office at the beginning of the Progressive Era and, in time, came to symbolize the nation's reform impulse, and reform was indeed in the air. At the heart of these reforms were Roosevelt's anti-trust actions against the near monopolies that controlled 40 percent of all manufacturing in the United States. Roosevelt was also instrumental in settling the coal strikes of 1902 through arbitration. Also during his administration, the Pure Food and Drug Act was passed. For all that these reforms promised, Roosevelt believed in orderly change as the only means to perfect and preserve the American system. It was Roosevelt the progressive, but also at heart conservative, who denounced muckrakers and socialists.

These reforms also helped to stabilize the rapidly-changing country, and strengthened the national government in the process. The new laws and commissions, while not threatening to capitalism, did place additional restraints on business activity. In spite of the reforms, and perhaps because of them, American capitalism and business thrived.

In Rockford the trend was toward growth and modernization. Rockford's population increased almost 50 percent, from 31,051 in 1900 to 45,501 in 1910, with a large portion of the increase due to immigrants arriving from Sweden and Italy. In 1903, a booklet titled *All Roads Lead to Rockford,* published by the Rockford Merchants' and Business Men's Association, encouraged industries and individuals to relocate in Rockford, going so far as to set up a committee "to entertain visiting delegates and excursionists." The Rock River was described as "a broad majestic stream," the tree-lined avenues as "veritable parks," and the manufacturing plants in the Water Power District as "a city by themselves, as they employ 10,000 people." No doubt this latter number was somewhat exaggerated, but the newly incorporated Woodward Governor Company with twelve employees in 1902, was indeed a small splash in that "broad majestic stream."[2]

The casual observer of Rockford found similar evidence of modernization. There were twenty-two miles of street railways that reached "every nook and corner of the city"; four river excursion steamers, the largest of which, *The Illinois,* had a capacity of 1,000 people; thirty miles of macadam streets and three of asphalt; and a water supply with a capacity of 9,000,000 gallons per day. The water supply came from nine artesian wells, not the Rock River. This accounted for the water's exceptional purity. The city also had two hospitals, one opera house, twenty-six churches, sixteen ward schools, one high school, Rockford College ("a school for ladies"), and Brown's Business College. The Rockford Baseball Association sponsored a team in the Three I League (Illinois, Iowa, and Indiana) and had won the league championship in 1902. The ball park had a capacity of 2,250 and was twelve minutes from the business district by the electric street railway. Four railroads passed through Rockford, and two separate interurban electric railroads connected Rockford to Belvidere on the east and Beloit and Janesville, Wisconsin, on the north.[3]

All this modernization partially hides the legacy of the nineteenth century's tremendous influence in Rockford. The city's first automobile had arrived in 1898. It was owned by William Barnes, Jr., son of one of the founders of the W. F. and John Barnes Company. By 1908, Rockford had 216 automobiles to run on its macadam roads. However, even at that time automobiles continued to frighten the horses and anger the owners of the horse drawn wagons still common throughout Rockford. These motorized "devil wagons" remained unpopular, and many

Elmer and Mary Woodward outside their home at 817 Haskell Avenue, Rockford.

people thought they were better suited to a rich man's whims or for racing than for the practical use of the average person. Blacksmiths and buggy whip salesmen still did a good business then, although a prudent person would have looked elsewhere for investment opportunities.[4]

While electricity dominated industrial and commercial Rockford, it was not so common among city householders, and it was practically nonexistent in the surrounding countryside. Consequently, kerosene lamp fuel was sold from wagons that still made the rounds of the residential districts on a daily basis.[5] Peddlers with their horse-drawn wagons were also common sights in the neighborhoods, providing door-to-door services for a myriad of farm produce and household products for the convenience of the housewife and her family. Less visionary people could only laugh along with many newspaper editors when conversation turned to flying machines. Most people simply thought that power-driven, heavier-than-air flying machines were a scientific impossibility. Balloon ascensions using hot-air bags were a county fair delight, but they seemed to serve no practical purpose except for observation of battlefields, a practice dating back to the Civil War. It is no wonder, then, that the Wright Brothers' achievement at Kitty Hawk in 1903 astounded the nation and the world. Within five years, planes could stay aloft for an hour or more and, by 1910, using bicycle wheels as landing gears, they could take off under their own power rather than from an inclined rail. Rockford itself would not see its first airplane until 1911, when a Curtiss biplane flown by Jimmie Ward and another plane flown by Beckwith Havens awed thousands of spectators by providing exhibition flights.[6]

Another innovation that affected Rockford just past the turn of the century was Thomas Edison's development of the motion picture. In 1893, he had perfected a way to take forty-five exposures a second on a continuous film and, by 1896, he had built the machine to project the images on a screen. The first movies lasted only about a minute-and-a-half, and they were sometimes used to lure crowds to temperance lectures. In Rockford, various kinds of lecturers brought films and finally, in 1904, a building on State Street was converted into a permanent theater named the Rockford Motion Picture Parlors. Admission was five cents except for the children's matinees on Saturday which cost three cents.[7]

In June 1903, President Theodore Roosevelt dedicated Rockford's Memorial Hall. The next year, Rockford built its first skyscraper—the six-storied Ashton Building. People thrilled to the first auto races held in a driving park north of the city. Also, the temperance movement made inroads into the collective psyche of the majority of the people of Rockford and, perhaps under the stronger influence of Rockford's women, the all-male electorate voted Rockford dry by 1908. Temperance had been one of Amos Woodward's goals for decades and, in 1904, he joined the Reverend Billy Sunday's temperance crusade by purchasing ten shares (at a dollar a share) of the capital stock of the Rockford Evangelistic Association.[8]

All of this bode well for the newly-incorporated Woodward Governor Company, not least because Elmer, nearing forty years of age, was in the full height of his inventive powers. The compensating governor of 1898, and Elmer's further improvement on it in 1901, propelled the business beyond Amos's abilities. He was over seventy years old, still head of the company, but he had left the task of improving governor technology to Elmer. Instead, Amos spent much of his time puttering around with other ideas, three of which had been patented. One was an oatmeal machine which he had invented in 1887. The others were an underfeed smoke-burning stove in 1906, and an exhibitor in 1910. None of these was commercially successful. Elmer himself also had patented some interesting non-governor inventions, including an orange sizer in 1892. Some sizers were sold, but Elmer and Amos stopped selling them after an unfortunate weather incident in Florida. After the sale, a severe frost killed most of Florida's orange crop. The growers had no need for the sizers, and had no money to ship them back to Rockford. Consequently, they were scrapped.[9]

There was another invention of Amos's wholly unrelated to governors that points up his keen interest in the benefits of fruit. This was a peach defuzzer. There is no record of a patent, but it must have been a good product with at least one highly satisfied customer. Long after Amos's death in 1919, and probably decades after the original sale, a Michigan customer wrote the Woodward Governor Company asking for a catalog and a price list for a new peach defuzzer. The original one worked so well, he noted, that Woodward must have improved it after all these years, and he would like to buy the new model.[10] There were, of course, no new models nor any of the originals left either.

Amos loved tinkering so much that, once he perfected some new device, he would start on something different, more or less oblivious to the fate of his newest invention or, for that matter, anything else that was in his way. As he worked with springs and gears, his hands would

Another Woodward patent, dated 1887, this one for an oatmeal machine.

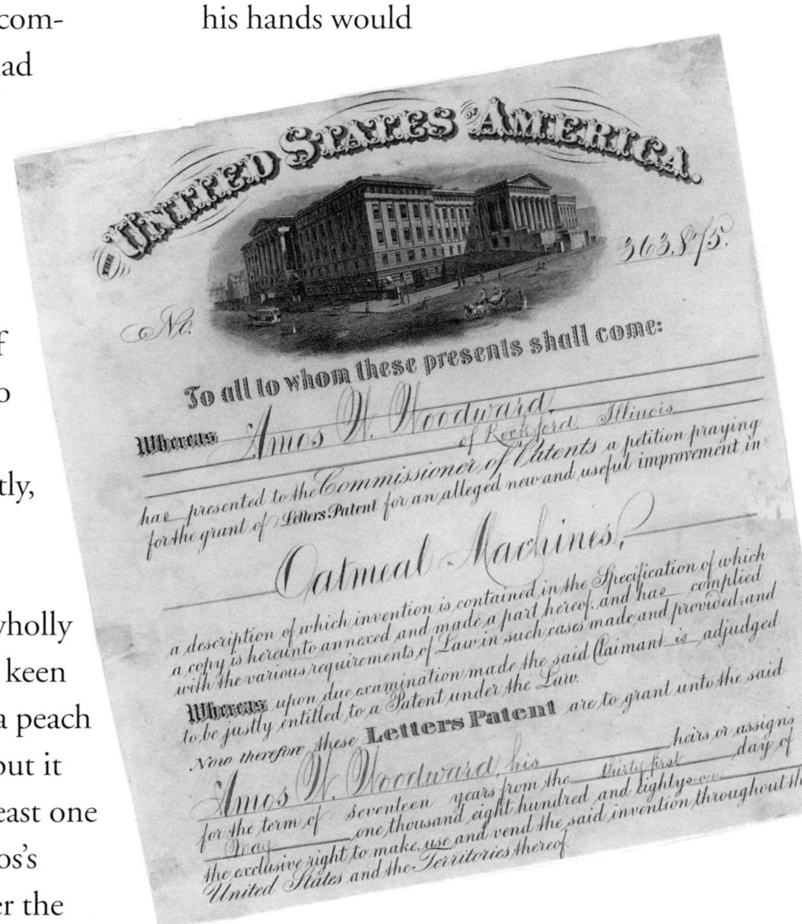
The United States of America.

No. 363,875.

To all to whom these presents shall come:

Whereas Amos W. Woodward of Rockford, Illinois has presented to the Commissioner of Patents a petition praying for the grant of Letters Patent for an alleged new and useful improvement in

Oatmeal Machines,

a description of which invention is contained in the Specification of which a copy is hereunto annexed and made a part hereof, and has complied with the various requirements of Law in such cases made and provided, and

Whereas upon due examination made the said Claimant is adjudged to be justly entitled to a Patent under the Law.

Now therefore these Letters Patent are to grant unto the said Amos W. Woodward his heirs or assigns for the term of seventeen years from the thirty first day of May one thousand eight hundred and eighty-seven the exclusive right to make, use and vend the said invention throughout the United States and the Territories thereof.

Woodward-Hess family Christmas gathering circa 1897. The Hess family was related to the Dunhams, whose daughter, Mary, married Elmer in 1887.

Front row: Walter Woodward (Elmer and Mary's son); Mary Sealy (Mrs. Amos) Woodward; Harold Woodward (Elmer and Mary's son); Amos, the family patriarch.

Second row: Cora Woodward (Amos and Mary's daughter); Dorothy Woodward (Elmer and Mary's daughter); Mary Dunham (Mrs. Elmer) Woodward.

Third row: Minnie Woodward (Amos and Mary's daughter); Ruth Hess; Elmer Woodward; Luther Hess.

Back row: Edith Hess; Birdie Fugate; Lissie Hess; Frank Hess.

pantomime gear rotations and other mechanical configurations. So engrossed would he become, that he often swept everything on his table into the nearest wastebasket, including the daily mail. Once Elmer caught on to what was happening, he rendered the eccentricity harmless by carefully combing Amos's wastebasket each day to retrieve checks and other valuable pieces of mail before the janitor burned them.[11]

As Amos's active part in the Woodward Governor Company declined, he concentrated more and more on the happiness of his two sons and two daughters, and continued his correspondence with his brother, Henry, in Maine. All of his children would become associated with the company in one capacity or another. Elmer was the anchor, the man most responsible for the company's successes for the forty years after 1900. Amos's second son, Franklin, who lived for a number of years in Chicago, eventually returned to Rockford and became associated with the company. His oldest daughter, Cora, worked as a stenographer for other companies, but she was living with and working for her father in 1903 when she tragically died at age thirty-three after an illness of a number of weeks.[12] Minnie, the youngest, married George Taylor and for a time lived in Ohio, but eventually returned to Rockford. Her son, George William (Bill) Taylor, had a long career with Woodward and died in 1978.

In the summer of 1907, Amos traveled to Maine to visit his brothers and sister. This family visit was a long-awaited tonic for him, and was much appreciated by his family. He impressed them with his generosity by placing new markers on the graves of their beloved, long-deceased parents. Henry, Amos's long-time correspondent since 1856, waxed eloquently over Amos's station in life when he wrote later: "Well, Walter [Amos], I rejoice in your success and prosperity in life. It is to me a matter of pleasure and pride."[13] By now, Amos, at age seventy-nine, was virtually retired except for his title and salary because Henry noted that "it must be pleasant for you now that you are able to take life easy with your children and grandchildren right about you."[14]

In 1909, Amos's eighty-eight-year-old brother, Franklin, died, and the following year his sister's husband died. Aware of his own mortality, Amos sent money to Henry and Priscilla, as well as some family pictures that included one with Elmer and his family and one with Minnie Taylor's son, Billy, on Amos's lap. Priscilla commented that forty-eight-year-old Elmer whom she had not seen since he was a little boy, was a "fine looking man."[15] Despite the family's close contact by letters over the years and Elmer's increasing travels on business sometimes to the Northeast, he apparently had never returned to Maine to see his uncles and aunts.

As the decade ended with Amos still in good health, he excitedly described to Henry the Woodward Governor Company's plans to erect its own building. Henry was impressed with the details about the new concrete building. However, while Henry was confident that this new building would lead to the company's continued success, he exaggerated by noting that it was big enough for 500 employees—a figure the company would not actually reach until World War II. But Henry was correct about two things: the company prospered throughout that first decade of the twentieth century, and Amos could "now sit down and let the boys run it and have the pocketing of the dividends."[16] The decision to declare dividends and the regular augmentation of these dividends underlines the company's financial successes.

When the decision to incorporate was made in 1902, business prospects were excellent. Elmer's new compensating governor was a good seller within two years after it was patented in 1898, and sales expanded so rapidly from the depressed decade of the 1890s that a corporate organization was warranted.

Amos Woodward posing with his grandson, George William (Bill) Taylor, son of Amos's daughter, Minnie. The photo dates from about 1910. Bill Taylor went on to a 44-year career at Woodward, serving as director from 1941 until his death in 1978.

On May 14, 1902, Elmer and his advisors filed legal documents with the Illinois Secretary of State to open books for subscriptions for the capital stock of the Woodward Governor Company. The capital stock was valued at $30,000, and 300 shares worth $100 each were outstanding. The stock was fully subscribed, with Amos and Elmer each subscribing to 125 shares and J. Charles Young to 50 shares. Little is known about Young, except that he was an employee proficient in mechanical engineering. Although he had a five-year contract with the company and was elected secretary and treasurer for the July 1, 1902 to June 30, 1903 fiscal year, Young was out of the company by the time the second annual meeting was held in

July 1903. During that first year, Young traveled around the country and in Canada to sell and service Woodward governors but, when the Board of Directors met in July 1903, he was absent. Elmer gave up his vice presidency and was elected secretary and treasurer, and Herbert M. (Bert) Woodward,[17] Amos's nephew, was elected vice president.[18] Since only stockholders could vote for directors and only directors could choose the company officers, it is probable that Young had been bought out and that his shares had been purchased by the Woodwards in 1903 with some of them going to Bert. The company became strictly a family corporation with that action.

When Amos and Elmer in effect sold the A. W. Woodward Company to the Woodward Governor Company in 1902, they valued it at $28,500. The assets included:[19]

Stock on Hand	$ 5,177.71
Patents, Patterns, and Good Will	10,000.00
Cash	95.18
Tools and Inventory	4,210.00
Accounts Receivable	12,100.00
	$31,583.69
Less Liabilities of	3,083.69
	$28,500.00

Thus, after thirty years of hard work, the value of the company had appreciated less than $1,000 per year. However, sales and profits accelerated after 1902 so that by 1910, when the company built its first building, the net worth of the company had nearly tripled.[20]

Elmer's compensating governors were more popular, more complex, more expensive, and more profitable than the non-compensating governors, and sales jumped from 104 units in 1902 to 131 in 1903. Favorable publicity also aided the company in its quest for sales. A Rockford newspaper noted in January 1903 that Woodward was going to double its capacity by leasing the space of the H. G. Dickinson Machine Shop, which adjoined it on the south side. The paper also noted that the company would be hiring extra help and then commented on the heretofore relative obscurity of the company:

> *While but few people in Rockford know much about the company, nevertheless it is one of the oldest firms in the city, and has advertised Rockford in more countries of the world than any other industry. The firm has shipped machines to almost every country on the globe, and is recognized as one of the three leading concerns of [its] kind in the United States.*[21]

While the article exaggerated the company's output at 200 governors per year, within two years that mark would be reached. Most importantly, this article and other articles in *The Iron Age* and *American Electrician* magazines featured and promoted the company and its products. These two trade journals, in fact, published lengthy articles complete with pictures of what they called the Woodward Friction Water Wheel Governor—which, of course, was Elmer's compensating governor.[22] Other trade journals that featured Woodward in 1903 included *Engineer, Street Railway Journal,* and *Electrical Review.* These articles were well-timed, and they undoubtedly helped to promote sales since the company at that time did not use conventional advertising.

The foundation of the business was its reputation, testimonials by satisfied customers, and a catalog that was periodically updated for new products or processes. The 1904 catalog was a very confidently written document that noted:

> *We realize that our product can only be sold on its merits, and consequently, we have endeavored to present these merits clearly, thereby allowing the judgment of the purchaser to determine the accuracy of our claim. We feel that the verdict cannot fail to be in our favor.*[23]

As before, Woodward guaranteed its governors virtually unconditionally, so long as the company's "instructions for installing and operating are properly followed."[24]

By 1904, the company was producing nine different governors—four of the compensating type and five of the non-compensating or standard type (see Table One).[25] The newest model, a horizontal compensating governor, was known as "size B." It was described as, "without doubt the simplest and most practical water wheel governor that is capable of filling the exacting requirements of electric light, power and traction service." It was the result of Elmer's 1901 patent, while the vertical compensating governors, sizes C and D, were based on Elmer's 1898 patent. The main advantage of the horizontal governor was that it could do heavier work than the vertical model. A smaller horizontal model, size F, was on the drawing boards and was also advertised and priced. Besides the more popular and quick-acting compensating governors, the company still sold the more basic standard governors used when slower gate movements were appropriate. Amos had patented the first one in 1870, and Amos and Elmer had improved this product and received patents for the improvements in 1890 and 1897.

The smallest of the standard governors (size 4) was practically obsolete by 1904, and it was not even listed in the 1910 catalog. Governor prices also remained stable during that six-year period. All the governors were the same price in 1910 as 1904, except the B horizontal compensating governor, which had dropped in price from $1,085 to $1,050.[26]

Often the company discounted the list prices for favored customers such as B. F. Groff, the sales agent for The James Leffel & Company. In 1906, Groff was able to buy a D governor from Woodward for $245 rather than list price of $375, and two C governors for $250 each instead of his usual discounted price of $280. This price was substantially below the catalog price of $525. Groff's rationale for the $250 figure was that his customer, the Patapsco

TABLE ONE

TYPE	WEIGHT	1904 PRICE
B Horizontal Compensating	1,850 lbs.	$1,085.00
C Vertical Compensating	750 lbs.	525.00
D Vertical Compensating	340 lbs.	375.00
F Horizontal Compensating	1,100 lbs.	750.00
No. 0 Standard Type	700 lbs.	300.00
No. 1 Standard Type	285 lbs.	187.50
No. 2 Standard Type	160 lbs.	150.00
No. 3 Standard Type	105 lbs.	112.50
No. 4 Standard Type	60 lbs.	Not listed

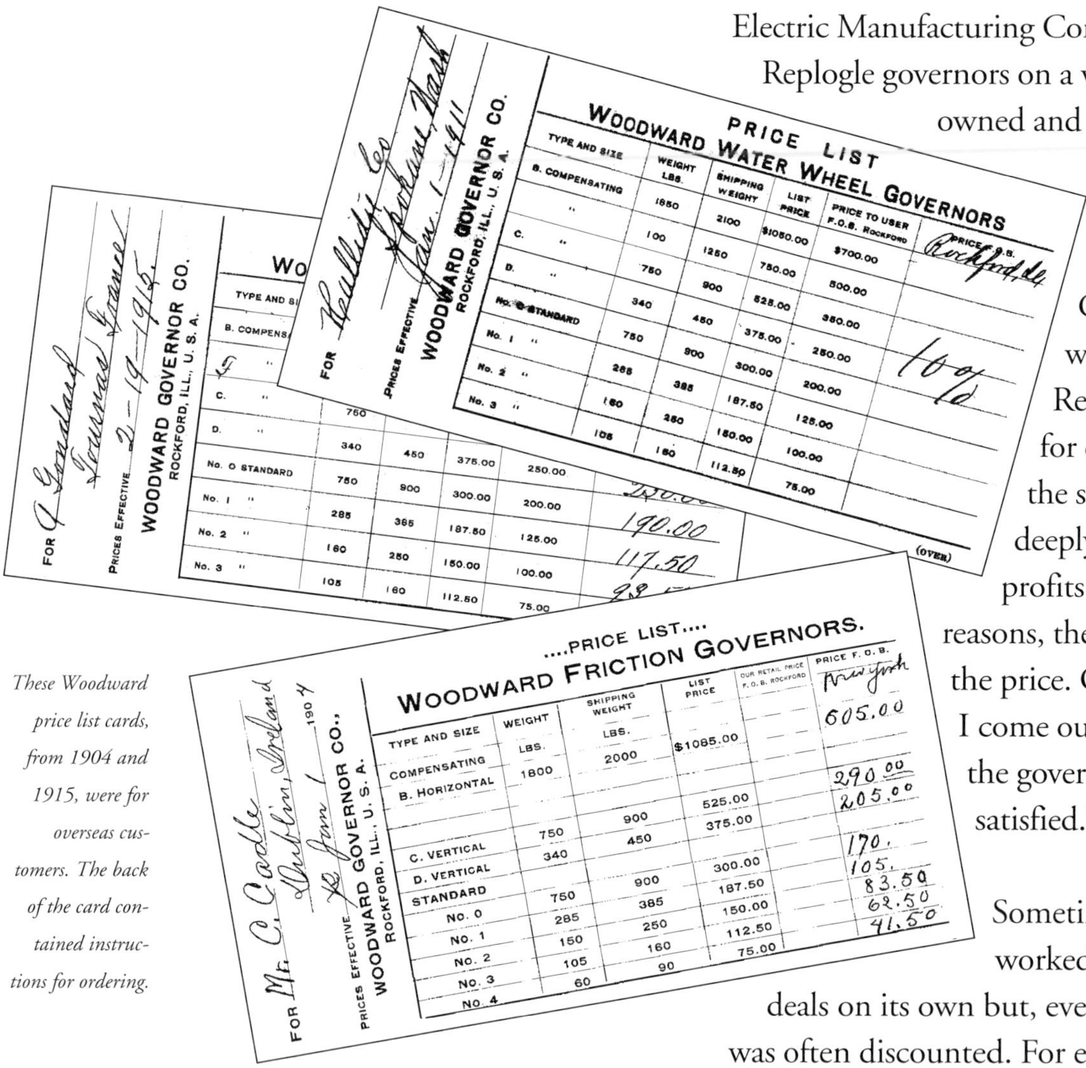

These Woodward price list cards, from 1904 and 1915, were for overseas customers. The back of the card contained instructions for ordering.

Electric Manufacturing Company, was using Replogle governors on a water power plant they owned and wanted to use the same make of governor in their new plant. According to Groff, Woodward would have to match Replogle's price of $250 for each governor to get the sale. This price ate deeply into any potential profits but, for competitive reasons, the company agreed to the price. Groff claimed that "if I come out even on the cost of the governors, I will have to be satisfied."[27]

Sometimes Woodward worked out more lucrative deals on its own but, even then, the list price was often discounted. For example, in 1907, the company offered to sell the city of Fergus Falls, Minnesota, a type C governor for $362 including delivery by train. This was well under the $525 list price, but $112 more than the deal Mr. Groff worked out in 1906. In 1908, however, Woodward sold a D governor to the city of Barton, Vermont, for $425, but included with that price delivery and installation by a Woodward man. The list price for a D governor was $375 and, for installations during this time, Woodward usually charged $6 a day for the installer plus his travel, room, and board. Therefore, the price was a fair one, and much closer to Woodward's list price than many of the other sales.[28]

In addition to its price discounts, the company's commitment to fairness meant that it would not try to sell a customer a more expensive compensating governor if a standard governor would suit the customer's purposes. This was the case, in 1908, when the Deer Creek Water and Water Power Company of York, Pennsylvania, contemplated buying a D governor. In a letter to Deer Creek, Elmer set out alternatives based on other equipment the York company intended on buying, and concluded that instead of a D governor with a $375 list price, Deer Creek could get along nicely with a No. 3 standard governor that Woodward would furnish for $75 instead of the list price of $112.50.[29]

In all of these instances, regardless of the price and terms of sale, the Woodward guarantee was ironclad. As the proposal to Barton, Vermont, made clear, the governor was guaranteed to regulate the speed of the wheel—

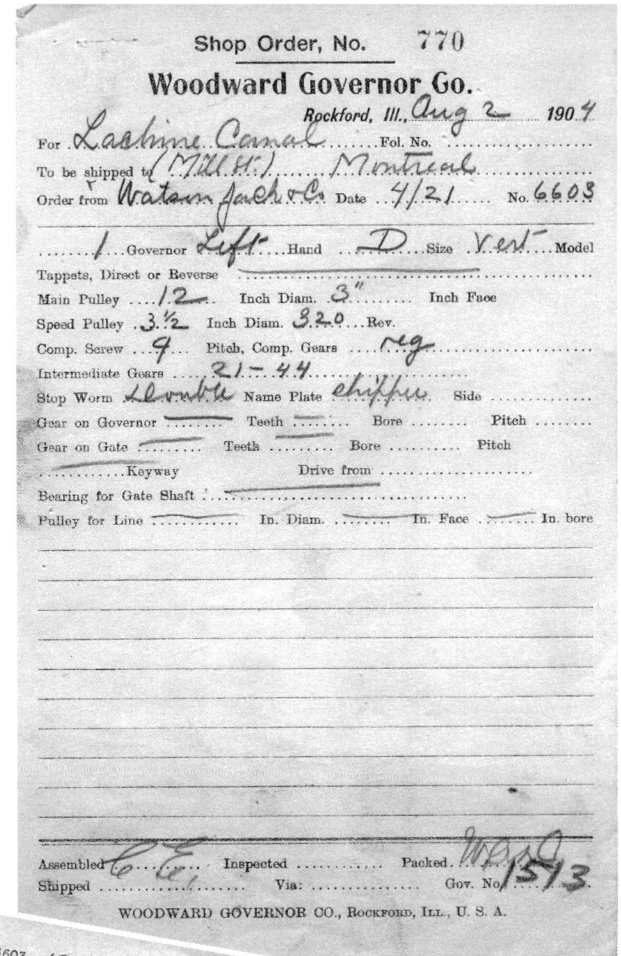

Shop Order, No. 770

Woodward Governor Co.

Rockford, Ill., Aug 2 1904

For Lachine Canal ....Fol. No. ....

To be shipped to (Mill H.) .... Montreal

Order from Watson Jack & Co. Date 4/21 .... No. 6603

1 ...Governor Left ...Hand D ...Size Vert ...Model

Tappets, Direct or Reverse ....

Main Pulley ....12.... Inch Diam. 3" .... Inch Face

Speed Pulley 3 1/2 Inch Diam. 320 ...Rev.

Comp. Screw ...4... Pitch, Comp. Gears ...reg...

Intermediate Gears ....21–44....

Stop Worm double Name Plate chipper Side ....

Gear on Governor .... Teeth .... Bore .... Pitch ....

Gear on Gate .... Teeth .... Bore .... Pitch

....Keyway Drive from ....

Bearing for Gate Shaft ....

Pulley for Line .... In. Diam. .... In. Face .... In. bore

Assembled C. E. .... Inspected .... Packed. W. A. G.

Shipped .... Via: .... Gov. No. 1513

WOODWARD GOVERNOR CO., ROCKFORD, ILL., U. S. A.

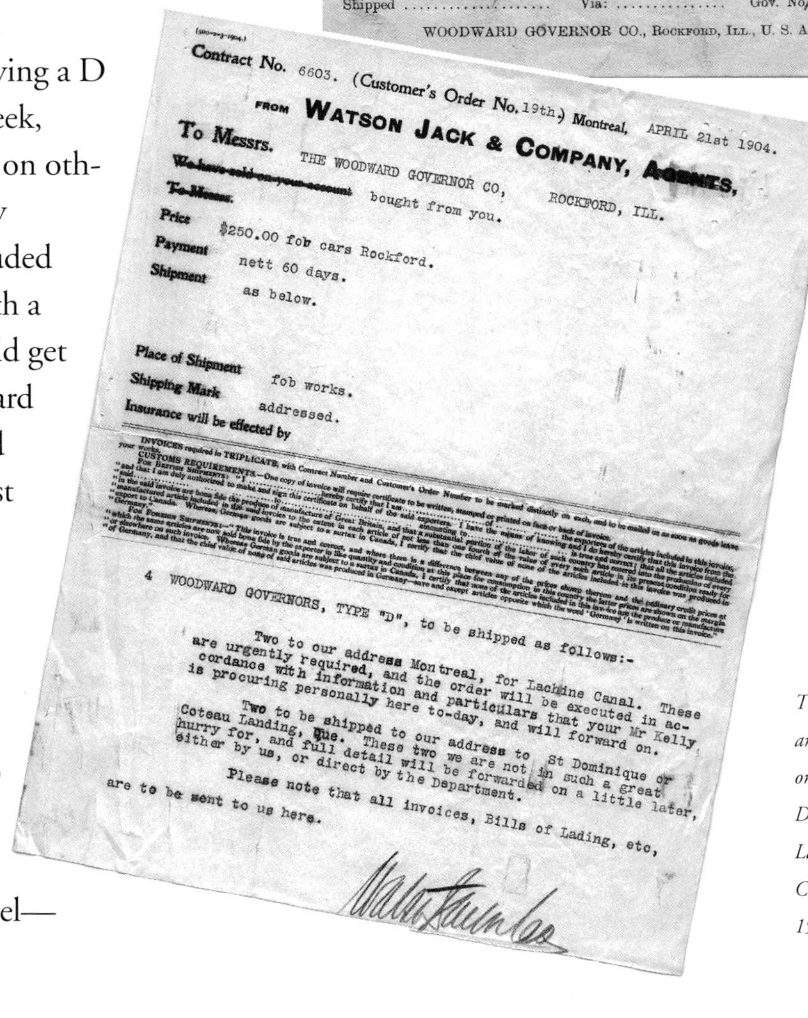

Contract No. 6603. (Customer's Order No. 19th.) Montreal, APRIL 21st 1904.

FROM **WATSON JACK & COMPANY, AGENTS,**

To Messrs. THE WOODWARD GOVERNOR CO, ROCKFORD, ILL.

~~We have sold on your account~~ ~~To Messrs.~~ bought from you.

Price $250.00 fob cars Rockford.

Payment nett 60 days.

Shipment as below.

Place of Shipment fob works.

Shipping Mark addressed.

Insurance will be effected by

4 WOODWARD GOVERNORS, TYPE "D", to be shipped as follows:-

Two to our address Montreal, for Lachine Canal. These are urgently required, and the order will be executed in accordance with information and particulars that your Mr Kelly is procuring personally here to-day, and will forward on.

Two to be shipped to our address to St Dominique or Coteau Landing, Que. These two we are not in such a great hurry for, and full detail will be forwarded on a little later, either by us, or direct by the Department.

Please note that all invoices, Bills of Lading, etc, are to be sent to us here.

The shop order and contract for one left-hand type D governor for the Lachine Canal in Canada, dated 1904.

Elmer Woodward installing the governor in the Lachine Canal powerhouse.

Woodward's domestic sales became more diverse and widespread. The company dealt with a wide variety of water wheel manufacturers as well as directly with various power companies and industries throughout the United States. Foreign sales were especially impressive. For example, sales to the Alessandro Calzoni Company of Bologna, Italy, had begun in 1898 when three governors were purchased. The number sold to Calzoni increased rapidly thereafter until they declined drastically after 1908 (see Table Three).[32]

to the entire satisfaction of your Engineer or Superintendent. We will further guarantee the governor and material furnished by us to be free from defects in material and workmanship, and if any parts are found to be defective or show a noticeable wear within one year, we will replace them free of charge.

We will allow ninety days after the governor has been installed for testing it, and if it does not prove to be all we claim for it, or does not meet the requirements in every way, it is to be boxed and delivered to your nearest railroad station subject to our order.[30]

With guarantees like this and with a superior product, it is no wonder that governor sales boomed from 1902 to 1910 (see Table Two).[31]

Sales to The James Leffel & Company in conjunction with Leffel's sales of its own water turbines to third parties remained strong during this period, and

While Calzoni was the biggest overseas customer, Neyret-Brenier and Company of Grenoble, France, ran a close second beginning in 1906 and, in 1907, for that year only, actually surpassed Calzoni by purchasing thirty-three governors to Calzoni's thirty-two. However, Neyret-Brenier dropped back to eleven in both 1908 and 1909 and, like Calzoni, bought only four in 1910. Even when sales to Calzoni and Neyret-Brenier declined after 1908, foreign sales remained a significant percentage of total sales. Countries other than Italy and France where large numbers of governors were sold included Canada, Mexico, and Finland, but governors were sold throughout the world. Larger numbers were sold in Europe, Central America, and South America than elsewhere, although some were purchased by companies in Japan, New Zealand, Formosa, Java (The Dutch East Indies), Jamaica, and Burma. In fact, it is not difficult to see that one reason for Woodward's prosperity in the first decade of the twentieth century was the successful marketing of its products to overseas customers (see Table Four).[33]

TABLE TWO

YEAR	GOVERNORS SOLD	YEAR	GOVERNORS SOLD
1902	104	1907	241
1903	131	1908	191
1904	161	1909	228
1905	201	1910	165
1906	242		

TABLE THREE

YEAR	SALES TO CALZONI COMPARED TO TOTAL GOVERNORS SOLD	YEAR	SALES TO CALZONI COMPARED TO TOTAL GOVERNORS SOLD
1899	8 of 36 (22%)	1905	32 of 201 (16%)
1900	17 of 75 (23%)	1906	39 of 242 (16%)
1901	29 of 88 (33%)	1907	32 of 241 (13%)
1902	19 of 104 (18%)	1908	34 of 191 (18%)
1903	15 of 131 (11%)	1909	18 of 228 (8%)
1904	24 of 161 (15%)	1910	4 of 165 (2%)

TABLE FOUR

YEAR	FOREIGN SALES	TOTAL SALES	% OF FOREIGN SALES
1900	24	75	32%
1901	36	88	41%
1902	42	104	40%
1903	39	131	30%
1904	79	161	49%
1905	75	201	37%
1906	88	242	36%
1907	91	241	38%
1908	75	191	40%
1909	48	228	21%
1910	34	165	21%

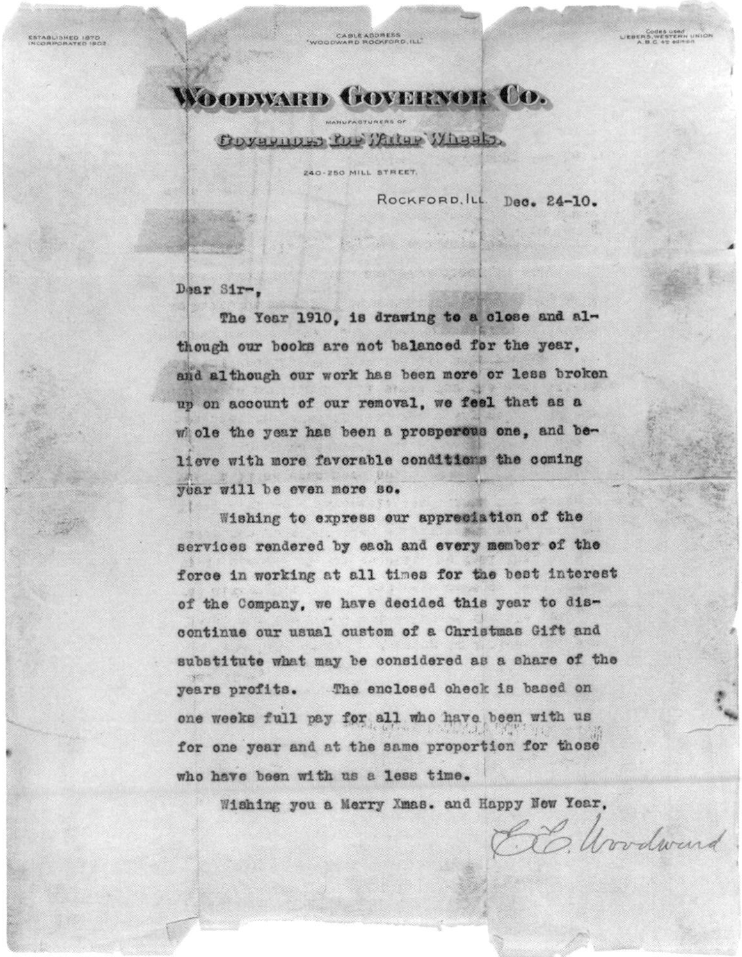

ESTABLISHED 1870
INCORPORATED 1902

CABLE ADDRESS
"WOODWARD ROCKFORD, ILL"

Codes used
LIEBERS, WESTERN UNION
A.B.C. 4th edition

WOODWARD GOVERNOR CO.

MANUFACTURERS OF

Governors for Water Wheels.

240-250 MILL STREET.

ROCKFORD, ILL. Dec. 24-10.

Dear Sir-,

The Year 1910, is drawing to a close and although our books are not balanced for the year, and although our work has been more or less broken up on account of our removal, we feel that as a whole the year has been a prosperous one, and believe with more favorable conditions the coming year will be even more so.

Wishing to express our appreciation of the services rendered by each and every member of the force in working at all times for the best interest of the Company, we have decided this year to discontinue our usual custom of a Christmas Gift and substitute what may be considered as a share of the years profits. The enclosed check is based on one weeks full pay for all who have been with us for one year and at the same proportion for those who have been with us a less time.

Wishing you a Merry Xmas. and Happy New Year,

E.E. Woodward

Elmer Woodward's 1910 Christmas letter to company employees. It introduces the word "member" and the profit-sharing concept that would form an important element of the company's philosophy.

This was a heady decade for the company, and it must have been exciting to see profits significantly increase for the first time in its history. The greater portion of the profits was reinvested into the business for new tools or plant improvement, or were added to the company's surplus. However, beginning in August 1903, on Elmer's motion as secretary, treasurer and general manager, the company began to pay irregular but increasingly larger dividends. The first dividend was 10 percent on the capital stock, which meant a cash payment of 10 percent of the $30,000 capitalization, or $3,000. Since there were 300 shares in the company, each share received $10. At this time, Amos and Elmer each owned 125 shares, so each received $1,250 in cash.[34] This was a stark contrast to the dark days of the 1890s, when the company was barely surviving, and Amos's and Elmer's yearly salaries had probably been below this first dividend. Three decades of hard work were finally paying off, making Amos's declining years comfortable.

That first dividend payment for the 1903-04 fiscal year was modest compared to the dividends paid out in the next three fiscal years. As total governor sales topped the 200 mark in 1905, and continued higher still in 1906 and 1907, higher profits were followed by higher dividends. Fifteen percent dividends were declared twice in fiscal 1904-05, and in 1905-06 two dividends totaling 35 percent were declared. Both of these years were topped by 1906-07, when three dividends totaling 55 percent were declared.[35]

Dividends remained high—at the 35 percent level—for fiscal 1907-08 then dropped to 10 percent in fiscal 1908-09 and were omitted in fiscal 1909-10. The reason for this was the company officers' decision to expand the company again by constructing a new facility on Mill Street, and to give each company member a Christmas bonus of one week's pay in 1910.[36] This was the first recorded instance of profit sharing by the company. Three decades later, it became an annual practice.

Elmer had begun preparations for the new building in 1907, when Woodward purchased two parcels of land in the Water Power District suitable for a manufacturing facility. The first parcel

After Woodward Governor Company left the Thompson building in 1910, it occupied this five-story building on Mill Street. The company remained here until 1941, when it moved into its new, and current, headquarters facility on North Second Street, Loves Park.

was purchased on February 11 for $1,800, and the larger second parcel was purchased on March 30 for $6,000. The second parcel included ownership of Rock River water rights (300 of the 20,000 shares) controlled by the Rockford Water Power Company.[37] To finance the new building, Woodward borrowed $30,000, which was secured by the issuance of a series of bonds and a mortgage on the real estate, manufacturing plant, property, and equipment of the company.[38] This conservative method of financing was in keeping with Elmer's cautious and methodical approach to business.

The construction of the new building began in June 1909, and was completed eight months later. Compared to the company's rented quarters, this new building was enormous and modern and showed how far the company had come in thirty years. In 1873, the A. W. Woodward Company was located in a small wooden building; in 1893, it had expanded into the three-story brick N. C. Thompson building; now, in 1910, it was moving into a five-story, reinforced concrete building with dimensions of 140 feet by 64 feet. The dominant architectural feature of each floor was 2,500 square feet of window space, which the company claimed provided "abundant light in any part of the building for the most delicate work," even on a cloudy day. Besides the main factory, the new complex included two smaller buildings which Woodward leased in 1910. In addition, a steam plant provided heat and power, and a hydroelectric

Elmer Woodward in 1910.

The fifth floor of the Mill Street plant in 1910. The man in the foreground is Dave Sandell.

powerhouse supplied electricity for lighting and power. Not surprisingly, the powerhouse was self-regulated by a Woodward type D compensating governor that maintained a constant voltage within 1 percent of the norm.[39] The new plant was a matter of pride to the company officers. In 1910, as a way to promote themselves and their product, they began a short-lived company publication called Power Control. Only three issues were published, two in 1910 and one in 1911, but the publication served as a prototype for company publications that were revived in the 1920s and continue today.

The first issue of Power Control described how to control power, especially water power. It promoted the company's own products as the solution for manufacturers, power plant operators, and others attempting to harness water power. Over 2,000 Woodward governors were in satisfactory service by 1910, and the company claimed to make governors "adapted to every need and condition." In calling on potential customers to let Woodward "help them to a right decision," Elmer reiterated a long-held personal philosophy on business that has remained a company tradition. The company was not in business just to accumulate property and money, because every dollar invested in the business represented—

> *an equal or greater value realized by the users of Woodward governors. The real success of any business is measured by the service it renders to its patrons, and that is the only brand of success worth talking or writing about. We mean to measure our future success by the same standard...*[40]

This philosophy certainly differentiated Woodward from mainstream American companies during the Progressive Era. Whereas they sought wealth, status, and power, the Woodward Governor Company was in business to provide service. At the same time, the philosophy at Woodward was also a contributing factor to the company's self satisfaction and conscious decision to stay relatively small and intimate. This intimacy certainly promoted Elmer's emphasis on the dignity of the people who worked for him and on the dignity of people whose vocation was manufacturing.[41] Thirty years later, when the company expanded again to become much larger, concern for promoting the dignity of the worker members of the company would still be in the forefront of the decisions made by management.

The Impact of World War I

1910-1919

In 1917, George M. Cohan wrote the words and music to "Over There," a rousing World War I song that symbolized thc idealism of the American people as the United States entered the war. After three years of neutrality, while the European powers waged the bloodiest conflict in history, the Yanks were finally over there, and the country would never be the same again. To the small Woodward Governor Company, World War I was to have two notable effects. First, the company would temporarily lose a portion of its export business, especially to Europe. And, a young army officer named Irl Martin, who was stationed at Camp Grant near Rockford, had a blind date with Elmer Woodward's daughter, Dorothy Carol. The company's foreign business began to rebound in 1919, after the war. In that same year, Irl returned from France and married Dorothy. Irl's entrance into the family would, in time, dramatically affect the Woodward Governor Company, for in 1921 he began a fifty-five year association with the company that would see him start out at the bottom before finishing at the top.

Irl and his future wife, Dorothy Woodward, on the porch of Elmer and Mary's home at 817 Haskell Avenue, taken about 1918.

But, in 1910, when the Progressive Movement was at its height and most of the world was at peace, Elmer Woodward was looking to his sons for the company's future leadership. He had two sons, Walter T. Woodward, then twenty-one, and Harold S. (Doc) Woodward, then eighteen. Although they did work for the company, both were to be overshadowed by Irl Martin after 1921.

Domestically, the most shocking political news in 1910 was the split between William Howard

Taft and his mentor, former President Theodore Roosevelt. Roosevelt had returned from a hunting trip to Africa incensed at what he saw as Taft's conservatism. Despite attempts at reconciliation, the split between them grew wider, and Roosevelt ran against Taft for the Republican nomination for President in 1912. Failing in this, Roosevelt ran as a Bull Moose Progressive in opposition to Taft and Woodrow Wilson, a Democrat. This Republican split practically guaranteed Wilson's election; he easily won the electoral college vote, although he received only 41.9 percent of the popular vote.

Under Wilson, a former president of Princeton University and governor of New Jersey, progressive legislation accelerated until World War I interceded. In 1913, the popular election of senators (previously elected by state legislatures) was confirmed by the Seventeenth Amendment, and the Federal Reserve Act, which created a government-controlled, decentralized banking system, was passed by Congress. In 1914, the Federal Trade Commission was set up to prevent unfair methods of competition in interstate commerce, and the Clayton Anti-Trust Act, which strengthened the Sherman Anti-Trust Act, was passed by Congress.

The war affected Rockford in a number of ways. War orders helped make the city prosperous. Rockford's citizens, concerned for war refugees and orphans, were generous in their philanthropic aid. War fever followed the U.S. declaration of war, and a "loyalty day" was held on April 7, 1917. A rejection of all things German soon followed, and people began to eat "liberty cabbage" instead of sauerkraut, the German language was banned in public schools, and Berlin Avenue's name was changed to Rockford Avenue.[1]

Despite his earlier pledge of neutrality, once President Wilson sent American troops to Europe, the country stood almost as one behind him. Entering the war seemed the right thing to do. The fighting had settled into a stalemate. America's entrance, it was hoped, would break the deadlock.

Rockford's patriotic support for the war included practical measures. Civic and business leaders were actively petitioning the War Department in the hopes of acquiring an army camp. For a site, they offered a large tract of land just south of Rockford. Finally, on June 7, 1917, Secretary of War Newton Baker accepted the site, and construction of Camp Ulysses S. Grant was begun in late June. Within a few months some 7,600 workmen had erected over 1,100 buildings to provide the support services necessary to house, feed, and clothe the 50,000 men that the camp was designed to accommodate.[2]

As hoped, Rockford boomed. Thousands of civilians flocked to Rockford to work on the project or in the auxiliary services that supported the great influx of soldiers. Restaurants, theaters, hotels, and retail stores flourished for the duration of the war. In general, the mix of thousands of soldiers and the city residents was cordial and trouble-free. But there were some tensions. Since Rockford was legally a dry town, it does not take much of an imagination to realize how difficult a job it was for the military police and the local authorities to enforce the ordinance. Young, footloose soldiers far from home, cooped up on a military base, took what opportunity they could to find more than the law allowed. And they found it in contraband liquor and imported women, who provided the kinds of illicit companionship the more conservative locals deeply frowned upon.[3]

Camp Grant was a central debarkation point for thousands of American soldiers sent off to fight, and sometimes die, on the battlefields of France. And then, during the fall and winter of 1918-1919, Camp Grant was not a safe place to be either, when the great influenza epidemic killed over 1,000 men at the camp and some hundreds of the residents of Rockford.[4] Irl Martin had left Camp Grant before the epidemic hit, and was stationed in a noncombat area near the western front in France.

During this decade, Rockford was a steadily growing city, adding an average of 2,000 people per year after 1910. By 1920, over 65,000 people lived there. Industrial activity also grew rapidly. In 1914, Rockford had 265 industries that employed 10,000 persons and paid $6,645,000 in wages. Five years later, these numbers had grown to 312 industries that employed 15,000 persons and paid $18,379,000 in wages. These figures continued to dwarf the small but successful Woodward Governor Company which, in 1917, had twenty-four employees who were paid $31,453.30, including a top salary of $5,040 to Elmer and a second high salary of $3,060 to Robert Kelley. Most of the factory hands earned between $900 and $1,200 a year

Irl (far right) as a second lieutenant in France, taken in 1918 or 1919.

for the normal sixty-hour work week. The president and patriarch, Amos Woodward, no longer received a salary but lived nicely on his stock dividends.[5]

A number of people in key company positions had already been long-standing employees by 1910. Robert Kelley had worked for Woodward since January 1, 1904 and, by 1910, he was the assistant chief engineer (subordinate to Elmer Woodward, the chief engineer) and senior salesman. Before coming to Woodward, Kelley had acquired ten years of experience in the manufacture and installation of water turbines for the Dayton Globe Iron Works Company in Ohio, and the Hamilton Manufacturing Company of Peterboro, Ontario. His principal work in 1910 was to design governors and supervise their installation on turbine water wheels.[6] He remained with the company for almost another thirty years, always in leadership positions and usually as the main contact for customers. At various times, he was purchasing agent, director, and company officer. He was virtually retired by 1938 but remained a stockholder until he was bought out by Elmer Woodward in 1940 when the company split its stock and increased its capitalization.[7]

Another employee with a longer career at Woodward than Kelley's was Cyrus (Cy) Erickson. He had joined the company in 1901 when he was thirty-six years old. He was an "old time" mechanic, superbly skilled in a myriad of tasks. He later became foreman of the mechanical governor department, where he stayed until he retired at age seventy-five in the early 1940s when the company moved to its North Second Street plant.[8]

One of the company's more remarkable employees at this time was Estelle (Stella) Milne, whom Amos had interviewed on December 31, 1901. She began work on January 2, 1902, and was to remain a company member for the next fifty years. For most of the first quarter century, she was the only woman in the company. She had come directly from Brown's Business College to Woodward. She began by handling all the office work, including sweeping the floor and cleaning off the iron dust that blew in through an open archway from the machine shop. She also prepared blueprints and carried the cash payroll from a nearby bank to the plant on paydays.[9] Before she retired in 1952, she served as assistant secretary, and she was also a stockholder for years. For much of her tenure, she had to deal on a daily basis with Irl Martin, who believed that the proper place for a woman was in the home. Perhaps he made Stella the exception to his rule, but since she never married, the full force of Irl's rule never got tested.

But the most important man in the company continued to be Elmer Woodward. He had joined the firm in 1877 and, after thirty-three years of increasing responsibility and success, he was still inquisitive, understanding, and innovative. His invention of the compensating mechanical governor remained a big success in 1910, but he kept up with his experiments, so that by 1911 he had perfected an even better governor—the Woodward V.R. (Vertical Relay) Oil Pressure Governor. This hydraulic governor operated faster than the mechanical governor. It could fully open or close the turbine gates in one second, five times faster than the compensating governors. But speed was not its only virtue. According to a company bulletin, accuracy and smoothness in the control of load changes were just as important. "The movements of the governor appear to the observer to be simultaneous with the changes of load, and there is absolutely no movement when the load and head are steady." The hydraulic governor's main selling points were that it was more durable, more accurate, more reliable, as well as faster, cleaner, and simpler. It also came in a variety of sizes.[10] It was a genuine breakthrough, and it quickly became the company's biggest seller. The first installation went into the powerhouse of the Rockford Iron Works in 1911.[11]

Although the number of governors sold declined in the 1911-1915 period, the company remained profitable and, when sales began to accelerate again between 1916 and 1919, the company thrived (see Table Five).[12] Foreign orders declined early in the decade primarily because two of Woodward's leading customers in the first decade of the twentieth century, Calzoni of Italy and Neyret-Brenier of France, stopped buying Woodward governors after 1912. At the onset of World War I in the summer of 1914, sales to the European continent were made more difficult when German submarines began to threaten all neutral cargoes. In 1915, the first full year of the war, none of the thirteen machines sold that year outside the U.S. went to the European continent. Four went to the Jas. Gordon Company in London, a regular customer, six went to various companies in Canada, and the other three were sold in Latin America. The following year, 1916, was not much better, when only twenty-seven machines were sold to foreigners—including five to Europe. All other sales were confined to the Western Hemisphere, in either Canada or Latin America. In 1917, as the war grew worse, only one of the seventeen machines sold overseas that year went to the continent of Europe, and England did not purchase any. Then, in 1918, sales to England and especially France increased. Six governors went to France, primarily for the national gunpowder works. This war-related business was matched at home by the sale of ten governors to the Rock Island Arsenal.[13]

The Type VR Governor, as shown in the 1923 "Net Price List" catalog.

TYPE V. R.

Vertical Relay Valve Oil Pressure Governor

Built in all sizes up 30,000 Foot Pounds Capacity.

Designed for connection to gate shafts turning 90 degrees or more to open gates full.

WOODWARD GOVERNOR COMPANY
ROCKFORD, ILLINOIS, U. S. A.

TABLE FIVE

YEAR	FOREIGN SALES	TOTAL SALES	% OF FOREIGN SALES
1911	44	164	27%
1912	27	136	20%
1913	31	153	20%
1914	17	119	14%
1915	13	117	11%
1916	27	171	16%
1917	18	175	10%
1918	22	191	12%
1919	34	199	17%
1920	35	248	14%

Compared to the previous decade, when foreign sales were usually between 30 percent and 40 percent of total sales, the decline was drastic. Fortunately, through frugality and the new hydraulic governor, the company remained profitable, and its net worth increased each year.

The earliest company document describing its assets and liabilities that has survived is a statement given to R. G. Dun & Company on January 1, 1907.[14] At that time Woodward's net worth was almost $66,000. In the five years since incorporation in 1902 at $30,000, the company had more than doubled in value. Two years later, net worth was up to $77,000. Since profits were good, net worth continued to rise during 1909-1910, when the new plant was built, even though the company borrowed money for its construction. By 1912, net worth was over $117,000. It continued to increase, although more slowly, throughout the teen years, until it dropped in 1919 due to a 60 percent dividend paid that year (see Table Six).[15]

TABLE SIX

YEAR		AMOUNT OF ASSETS OVER LIABILITIES
February	1912	$117,031
February	1913	122,594
January	1914	133,386
January	1915	137,167
January	1916	144,473
February	1917	152,297
January	1918	175,444
December	1918	193,730
December	1919	188,205

Business was generally good during these years, although in some years only marginally so. Therefore, especially in the early teen years, dividends were modest compared to the first decade of the twentieth century. They began to increase in 1916. During World War I and immediately afterwards, they were especially high. Moreover, both in mid-1918 and mid-1919 Woodward paid its members extra compensation as a bonus.[16]

The company operated on a fiscal year from July 1 to June 30. However, a comparison between calendar year net income and calendar year dividends more clearly shows the relationship between income and dividends (see Table Seven).[17]

The company was very fortunate that Elmer invented the hydraulic (oil pressure) governor when he did—for, in the years after 1910, sales of the obsolete mechanical non-compensating governors were minuscule. Even sales of the compensating governors were in decline because they seemed slow and old-fashioned compared to the rapid and easy-to-use hydraulic governors. The hydraulic governor debuted in 1911, with meager sales of $297 out of total sales of $26,965. By the end of 1912, hydraulic governors worth $7,059 had been sold, putting the new product second only to the $10,009 that the size D vertical compensating governor produced in sales. The next year, when hydraulic pumps were

TABLE SEVEN

YEAR	NET INCOME	DIVIDENDS AS A PERCENT OF $30,000 CAPITALIZATION	ACTUAL COST TO COMPANY
1909	$22,626	No dividends	$ —
1910	14,297	10.0%	3,000
1911	5,699	10.0%	3,000
1912	5,666	No dividends	—
1913	13,113	7.5%	2,250
1914	6,372	10.0%	3,000
1915	10,603	No dividends	—
1916	14,075	40.0%	12,000
1917	29,448	15.0%	4,500
1918	Not Available	12.0% + 6.5% wage bonus	3,600 +
1919	Not Available	60.0% + extra	18,000 +
		compensation of $2,969 to members	2,969

Amos was dependent on these sporadic dividends for his livelihood, since during these years he no longer drew a salary, even though he remained president. He did not live to see the 60 percent dividend of 1919, and would probably have been amazed to see how the dividends skyrocketed in the 1920s.

added to hydraulic governor sales, the total was $10,500. D governor sales that year totaled $13,421. By the end of 1914, the contest was over and hydraulic governors and pumps were the clear winners. Their sales of $31,900 were three times the D governor sales of $10,038. The gap widened by the end of 1915—$37,800 to $8,926, and it undoubtedly continued in the same direction for the rest of the decade.[18] Hydraulic governors were the company's savior for the teen years and its major hope for the future.

The company continued to operate much like an extended family during this decade. Elmer was the boss and the chief engineer, but his authority was benevolent and modest. He negotiated conflicts and treated people fairly. Thus, the loyal and talented individuals he gathered around him became members of an extended family. Some of them, like Estelle Milne, Robert Kelley, and Cy Erickson, had already spent almost two decades with Woodward. Others, like David and Fred Sandell, George Sorensen, and Ward Ring, who had joined the company later, would stay on for the rest of their working lives and, except for Sorensen, who died in 1942, would not retire until a decade or more after World War II. Consequently, this latter group witnessed the next major transformation of the company as it expanded again in the World War II years.

George Sorensen was hired in March 1911, when he was almost thirty years old, to do drafting and design work. Prior to this, he had worked for a Rockford furniture factory as a carving machine operator. He had also worked for Ingersoll Milling Machine Company and W. F. and John Barnes. During these years, he studied engineering through a correspondence school—an education he described as "half-baked." Nevertheless, it did help him secure a job at Woodward where he was taught the governor business by the acknowledged expert, Elmer Woodward. The lack of a formal engineering education was not a detriment to George Sorensen any more than it was to Elmer.

As George learned more about the compensating governors and the new hydraulic governors, he began to go on service trips, first with Elmer and then on his own. He was involved in the engineering that helped develop the hydraulic governor, and his ink tracings on vellum were practically works of art. He quickly made himself valuable, perhaps indispensable, since by the 1920s he owned 5 of the 300 shares of stock in circulation.[19]

When George died in 1942 at age sixty-one, after complications from surgery, he had served the company for thirty-one years. In the process, he became the designer whose drawings incorporated Elmer's inventions and helped put them to practical use. He also served the company as vice president and later as treasurer.[20]

The Martins of Broken Bow, Nebraska, about 1910. Seated: parents Charles Stoner and Josephine Martin. Standing, left to right: Charles Alton, Day, Nell, Irl, and William.

Ward Ring began his career in 1913, just two years after Sorensen started. He was hired by Frank McHugh, the shop superintendent, for what was then considered good pay—50¢ per hour. Ward had worked for Barber-Colman, but a friend of his at a local foundry had recommended to Bert Woodward that Ward be hired. He started in production, but he also became a tool maker, tool designer, heat treater, metal pattern worker, and millwright. Eventually he worked up to machine shop foreman and shop superintendent. During the teens, when mechanical governors were still being made in large numbers, they were built for stock. Tolerances had become more exact over the years, but they still were not close, particularly in comparison to the last fifty years. In fact, Woodward had only a makeshift grinder, and it was used for both internal and external work.[21]

While Ward Ring had a long, distinguished career with Woodward, it was not continuous, for records show that he was not paid the bonus in June 1919 for "present employees...for services rendered during the calendar year 1918."[22] According to Ward, during World War I the American Federation of Labor called a strike of all the machinists in Rockford. Although Woodward Governor Company was not a union shop, some of the machinists, including Ward, walked off the job. The union assumed, wrongly it turned out, that the government, in need of war production, would force the plants to pay higher wages. The scheme failed. Ward hauled lumber and helped construct Camp

Grant in the interim. Eventually, he and some other machinists went to Elmer and apologized. Their jobs were reinstated, without recriminations or hard feelings from Elmer.[23] There were other departures from the company for other reasons between 1917 and 1919. Of the twenty-four people employed in 1917, only fifteen were with the company throughout the entire year of 1918 and were, therefore, eligible to receive the retroactive bonus in 1919.[24]

Although the war increased Woodward's business, a longer-lasting benefit to the company arrived at Camp Grant in the person of Second Lieutenant Irl C. Martin. Irl was born in Broken Bow, Nebraska, on December 12, 1895. He, his brothers, and one sister had loving and successful parents who believed that their success had come from hard work. Naturally they expected their children to work hard too, and they saw to it that they did. This work ethic had a lifelong effect on Irl and greatly influenced his thinking during the years that he helped transform the Woodward Governor Company into a unique corporation.

Although Irl's parents had very little formal schooling, they made up for it with an abundance of horse sense. Irl's formal education did outpace his parents' by a considerable margin, but he consistently reflected their views on the gratifications and rewards of hard work:

> *Too many youths finish high school or college with an attitude toward good honest work as a penalty rather than as a maturing and satisfaction-developing adventure. Some academic experience is desirable and, like strychnine, it can be helpful if taken in a proper amount but can be very enervating and destructive if indulged in beyond the individual capacity to make constructive use of it. Through a lifetime, education is about 10% academic and 90% practical experience at best and, in most instances, a 1% - 99% ratio would be more realistic and of greater benefit in general.*[25]

Irl's parents required that he and his siblings do chores "regularly and without alibi or compromise." Once, when he was eight, Irl thought to challenge his parents when he went to bed without doing a routine chore. His father found him out, rousted Irl out of his bed, and forced him to complete the chore. He did it, simmered all night for the indignity he suffered and, in the morning, informed his parents that he was not going to do chores anymore. In fact, he was going to leave home. His bluff was called immediately when his father and mother offered to pack his bag and some food to take on his trip "into the world." This was not the reaction he expected, and his rebellion was broken immediately in the full knowledge that to stay home meant his assigned chores would have to be done "punctually and properly."[26] He never forgot that incident, and it affected the way he perceived work the rest of his life.

Another economic lesson learned early came from his ambitious, but ill-fated, attempt to curry favor with his friends through loans and treats like ice cream sodas. He became quite popular for a time until his scheme fell apart at the seams. To finance his popularity, Irl had taken to selling out-of-town weekly papers for a commission, the money from which he used for loans and treats. Before he knew it, however, his friends were demanding more than he had money to give. Thus, he began to borrow from the portion of his sales that was supposed to be returned to the publishing company. Now Irl was in debt, and when his father found out that he had defaulted on payments, he made Irl work at additional money-making tasks until he paid off all that he owed the publishing company. In the meantime, his fair-weather friends abandoned him, since he was no longer supplying them with loans and treats. The experience taught Irl two lessons: friends can't be bought, and "there are few satisfactions as complete as being economically free and unbeholden."[27] Both of these maxims would affect the way he would later run the Woodward Governor Company.

But life for Irl was not all hard work. There was plenty of time to play. Irl's closest friend was Fred Humphrey, who lived across the street.[28] They were involved in all of the major sports: football, basketball, and baseball—and also played hockey and attempted bronc-riding. Their preteen baseball team was called the "Neversweats," and their fathers coached them and took them by wagon and team to neighboring towns to play games. By the time Irl and Fred were in high school, they were practically inseparable in their sporting activities. Then athletics came to an end for Irl when he broke his back playing football. He considered football to be the major cause of his hip joint problem, which led to hip replacement surgery when he was in his sixties.[29]

After Irl finished high school in 1913, he worked for the Custer Electric and Artificial Ice Company, which was owned by his father. His salary to run the power and ice plant, to read meters, and to sell ice was $2 a day, and he worked there until 1915. His father, meanwhile, had discouraged him from going to college, saying that if Irl wanted to go to college, it would have to be

Irl Martin played as a 135-pound fullback for Broken Bow (Nebraska) High School. The photo dates from about 1912.

with his own money, not his father's. To get around this, Irl talked his older brother into financing him and, in 1915, he headed east with plans to attend Rensselaer Polytechnic Institute in New York. He never got farther than Chicago, however, where he enrolled at Armour Institute of Technology. He was not a particularly serious student and claimed to have "wasted time for two years from 1915 to 1917." Then, when the United States entered World War I in 1917, Irl, having had experience in the National Guard, asked for a commission in the army. The army complied, and Irl was sent for training to the newly-established Camp Grant.[30]

While at Camp Grant, Irl had a blind date with Elmer's daughter, Dorothy Carol, with whom he fell immediately in love. For a long time he was the active pursuer, and she was the more reluctant partner. By the summer of 1918, however, just before his unit was ready to go to France, she assured him that she loved him, too.[31]

He was thrilled, but his patriotism also clashed with his personal hopes when he noted, "I have got to get away or there is going to be a marriage... Honey, if it wasn't for the disgrace of having to stay behind, I would pretty near be willing to stay, just to be here." Once he got to France he wrote, "I am sorry now that I didn't play the cave man stunt and marry you before I left."[32]

Irl spent two years in France doing boring and routine administrative jobs. After the war ended on November 11, 1918, he was especially eager to return home, but he knew that this was unlikely for many months. To compensate, in letters he poured out his love for Dorothy and shared his despair and loneliness with her, especially when her own letters didn't arrive regularly. He became apologetic and insecure but professed hope that they could get married "as soon as I set foot on American soil again."[33] In December, he sent her plans for the den of the house he was going to build for them. Just prior to Christmas he went to Paris and bought an engagement ring, which she accepted.[34]

In early 1919, while Irl was still in France, the patriarch of Dorothy's family, Amos Woodward, died at age eighty-nine years, eight months. He had lived a full life, had outlived both of his brothers, and was survived by his sister, his wife, and three of his children.[35] While he had not been a factor in the company for a number of years, his passing symbolized the end of the first era of the company's existence. Elmer would carry the company to new heights for the next twenty-one years. For nineteen of those years, he would be increasingly aided by Irl Martin, who was discharged from the army on May 5, 1919 and married Dorothy on September 15, 1919. The first generation was gone just as the third generation came upon the scene.

Irl (center, front) posing with high school friends in 1913. Fred Humphrey, Irl's best friend and sports teammate, is third from the left, between two women wearing sweaters.

The Prosperous 1920s

1920-1930

In time, Irl Martin would come to dominate the Woodward Governor Company, and his ideas, some of which were based on concepts learned from Elmer Woodward, would transform the company into a unique industrial organization. But all this was in the future. Meanwhile, after their marriage, Dorothy and Irl moved to Broken Bow, Nebraska, where Irl went to work again at his father's electric and artificial ice business. Irl was close to his parents and brothers and looked forward to his share of responsibility in the family-owned company, which was prosperous and well-run. Irl also liked the easy familiarity of Broken Bow, where life for an ambitious small-town businessman was good in the early 1920s. Dorothy, however, was unhappy living so far from her family in Rockford. Her opportunity to return came in 1921, when Elmer visited Irl and Dorothy. Noticing what a hard worker Irl was, Elmer invited Irl to join the Woodward Governor Company. There was probably a long-range motive to this invitation. Neither of Elmer's two sons had expressed much interest in the company, certainly not enough to make Elmer feel comfortable that he could some day turn it over to their direction. By offering Irl a job, Elmer hoped to solve this problem. When Irl accepted Elmer's offer, the news was not only good for Elmer, but for Dorothy as well. Now she could move back to Rockford. Irl literally started at the bottom in 1921, as a bolt threader but, within eight years, he was general manager and heir apparent.[1]

Elmer's judgment in offering Irl a job was as acute as his general entrepreneurial skills and inventiveness. In time, Elmer and Irl

The house where Irl Martin grew up in Broken Bow. When Irl and Dorothy married and moved to Broken Bow, a section of this house was physically detached and moved to an adjacent lot and served as their new home.

made a formidable and harmonious team because each had skills that complemented the other. Elmer was almost sixty in 1921, but his creative powers would continue unabated throughout the 1920s and 1930s. He seldom slowed down and never took a vacation. Irl was equally driven since, in childhood, his parents had impressed upon him the Horatio Alger ideal that hard work would lead to success. His ambition was boundless, and his industrial insight developed rapidly under Elmer's tutelage. Their relationship was based on mutual admiration and respect and, once Irl knew the business, his responsibilities increased rapidly. In later recollections, Irl remembered his early years with Elmer:

Elmer E. Woodward was a very unusual man. He was quiet, understanding and capable. He was a one-man business. He could handle any part of the business including investing, insurance and managing salesmen.... In the early phase he would visit a plant to take measurements then go back and manufacture a governor from scratch, install it, make it work and collect his money.[2]

Irl was in awe of Elmer, and even when he said that Elmer "was limited in scope because he had no greed to become big—only the best,"[3] he delivered an unabashed compliment. For Irl, being the best became an important theme when it came his time to run the company after Elmer's death in 1940. Although Elmer was reluctant to turn over management controls to others, he made an exception for Irl once he was satisfied that Irl knew the business. Therefore, after 1929, Elmer left the business details to Irl so he could concentrate on engineering. Despite this division of powers, Elmer and Irl's desks were adjacent, and Irl regularly briefed Elmer on business details. If Elmer "ever said 'No,' that was it."[4]

Irl and Elmer were kindred spirits when it came to working. If Elmer was working on an idea or project, time meant nothing to him. He would work as late as necessary, get a few hours sleep, and start over fresh early the next morning. Irl was the same way, and he expected as much from the other employees. "He was hard driving," one employee who remembers him said. "The company was his life, and he couldn't understand why it wasn't that way for everybody. He didn't care how many hours he put in—he didn't think anybody else should either." In his constant efforts to improve the company, he was so oblivious to time that his wife once had to send him a telegram inviting him home for dinner![5] Thus the company entered the decade of the 1920s with Elmer firmly in charge, and left it with his son-in-law a rising star.

When the United States entered World War I, it was a debtor nation. It came out of the war a creditor nation.[6] Its economic power was matched by its influence abroad. American aid to the Allies, combined with the large military force it had sent to fight the war, had been decisive factors for victory. But events also had a way of rolling along under their own momentum. President Wilson's hope of making the war an idealistic crusade to end all wars was not borne out. The Treaty of Versailles was not ratified by the United States Senate because of the fear it might infringe upon the freedom of action of the U. S. government. Moreover, the treaty made for an uneasy, vindictive peace, largely because of the huge penalties it imposed on Germany. It thus helped set the stage for a resurgence of German nationalism, the rise of Adolf Hitler, and the coming of the Second World War.

The war and its aftermath also had some dislocating effects on the economies of both the United States and the rest of the world. Europe was especially hard hit, and it never fully recovered in the decades between the wars. Wilson's idealism had also dissipated into national disillusionment with the less than satisfactory outcome of the Versailles Peace Conference, the rise of Bolshevik Russia, and the resulting Red Scare in the United States in 1919-1920. Nevertheless, at least two reforms given impetus by the war did continue into the immediate postwar period. The first was prohibition, an issue dear to the heart of Amos Woodward. The Prohibition Amendment was ratified in January 1919, only two months before Amos's death. The other was the Women's Suffrage Amendment, which was ratified in August 1920.

Changes in politics were also in the wind. By 1920, it was clear that a conservative Republican could defeat any Democrat for President. The men who would benefit from this were Senator Warren Harding, the compromise candidate for the Republican Party, and his vice presidential candidate, Calvin Coolidge. They easily defeated Governor James Cox and his running mate, Franklin Roosevelt. Harding waged a "front porch" campaign from which he called for a "return to normalcy"—a phrase which struck a responsive chord in the American people. Harding's victory introduced an era of conservatism that lasted throughout the 1920s. During this period, businesses and the philosophy of business growth reached the height of their influence in the United States.

Harding and Coolidge were unabashed supporters of the idea that business had made America great and that the best protection the national government could give business was the ability to pursue growth without major government restraints. Coolidge's view that "the business of America is business" reflected the prevailing belief that success came as the result of hard work, frugality, and honesty. What the 1920s possessed most of all was the confidence that capitalism could only improve the American economy.[7]

The economic trend was not always pointed upward in the 1919-1929 period, although that certainly was its principal direction. There was, right after World War I, a slight economic drop, but a business boom developed in the spring of 1919 that lasted into mid-1920, followed by a severe but short-lived economic recession. This lasted throughout 1921 and was caused primarily by declines in exports, government spending, and farm income. These last two continued to drop in the 1920s—the first by design of Harding's conservative fiscal policies, the latter because of a drop in the prices of basic commodities and the decline in foreign markets as the world recovered from World War I. The economy recovered again by 1922, and except for farmers, the rest of the 1920s were increasingly prosperous. Optimism was so high, in fact, that some people believed that poverty could be banished permanently, not by government action but by the efforts of business to provide high employment. The belief was based on expanding purchasing power of the American worker and on the encouragement of consumer spending through credit buying, which together would carry the economy and thus the people to new heights.

These national business values were certainly the kind that Elmer Woodward and Irl Martin approved. These values proved themselves not only in the company, but throughout Rockford. While the Woodward Governor Company remained small in the 1920s, it and Rockford were very prosperous. Rockford's population grew at an average of 2,000 people per year between 1910 and 1930, from 45,000 in 1910 to 65,000 in 1920 and 85,000 in 1930. One reason for this prosperity was Camp Grant. Its founding in 1917 had immensely helped Rockford's economy and population growth. The federal government had spent over $13,500,000 for the land, general con-

The Assembly area in the Mill Street plant, sometime in the 1920s. A type A actuator governor is shown in the right foreground.

struction, and the installation of utilities. During and immediately after World War I, the camp bought merchandise over the counter in Rockford stores at a rate of $100,000 or more a month. By the end of demobilization in 1919, more than 1,000,000 men had passed through Camp Grant. The money they had spent in Rockford was a boon to the local economy. Even the government-authorized dismantling and salvaging of much of Camp Grant in 1922 aided Rockford, because equipment and building material were sold at deep discounts, usually a few cents on the dollar.[8]

By the early 1920s, Rockford was so prosperous that the Chamber of Commerce was convinced that its manufacturing diversity practically guaranteed that long periods of unemployment could not occur. If any further evidence for this confidence was necessary, one had only to note the negligible effect the recession of 1921 had on Rockford. Manufacturing continued to expand throughout the 1920s, and the result by 1929 was especially impressive. In 1919, 14,922 people were employed in Rockford, and they produced goods valued at $74,919,000. Ten years later, 19,916 industrial workers were producing goods valued at $157,090,301.[9]

So much manufacturing activity encouraged expansion and the introduction of new businesses in Rockford. One such business was started by the Chappell Brothers, who manufactured Ken-L-Ration dog food and canned horse meat (considered a delicacy in Europe). Two new banks were founded—Security National and Commercial National. Other companies expanding during this time included three more banks, the National Lock Company, Rockford Furniture Company, and Sundstrand.[10] Orders also rapidly increased for Woodward during the 1920s and, to accommodate this, the company expanded into parts of its 1910 building that had previously been rented to other businesses.

Rockford's building boom reached its height in the mid-1920s. Residential construction peaked in 1925 when 641 homes valued at $2,541,400 were constructed. Two years later, buildings worth $6,550,873 were under construction, including the $2,750,000 eleven-story Faust Hotel, the Coronado Theatre, the Rockford Morning Star Building, the LaFayette Hotel, and the ten-story Talcott Building.[11]

Rockford's sights were not set on lofty economic statistics alone, however. The sky was, literally, the limit once the Lone Eagle, Charles Lindbergh, had successfully crossed the Atlantic Ocean from New York to Paris in May 1927. Interest in aviation naturally skyrocketed. The next year, Rockford made the national news when the *Rockford Daily Republic* promoted a goodwill flight from Rockford to Stockholm, Sweden. Two attempts were made in 1928. The first, on July 28, 1928, ended abruptly when the plane, "The Greater Rockford," overloaded with 400 gallons of gasoline and a tail full of souvenirs, crashed in a cornfield shortly after takeoff. The pilots, Bert "Fish" Hassell and Parker

"Shorty" Cramer, were uninjured, and the plane was repaired. On August 16, they took off again. After refueling in Cochrane, Ontario, they got as far as Greenland before making a forced landing on an ice cap. The pilots were rescued, but "The Greater Rockford" remained on the ice cap until 1968.[12]

But failure didn't easily dampen the optimism in Rockford as its industries continued to innovate and progress in the late 1920s. These good times were reflected at the Woodward Governor Company, where the 1920s was the most successful decade in its sixty-year history. These years saw rapid increases in retained earnings and in Elmer Woodward's accumulation of a measure of wealth through his salary and stock dividends from the company. Both retained earnings and Elmer's personal resources would help keep the company in operation and its workers employed despite the lack of business brought about by the Great Depression of the 1930s.

The staff continued to work sixty hours per week during the 1920s—a figure unchanged since the nineteenth century. Despite the long hours, the modest pay,[13] the old equipment, and the generally messy surroundings, Elmer Woodward remained a popular employer. Workers cherished his inventive mind, but they revered him most as a fellow worker and craftsman. He thoroughly enjoyed working on the floor among the machinists and disliked spending time on administrative work, which he often took home with him. In many ways he was married to his job, not so much for the money as for the old-fashioned satisfaction of making a superior product. He was always concerned with the quality of the product and with the company's reputation among its rivals and customers.

The superior line of hydraulic governors Elmer had developed in 1910 and 1911 was literally paying dividends by the late 1910s and early 1920s. By 1920, Elmer had also invented, designed, developed, and started manufacturing the first completely automatic hydraulic governor. This allowed remote control of a hydroelectric governing unit, and it became very popular in the 1920s. Production of Elmer's next breakthrough, the first successful motor-driven governor head for hydraulic governors, began in 1921. This quickly captured the market because it did away with the cumbersome and sometimes dangerous belt drives. By 1928, the switch to motor-driven governor heads was nearly complete. Motor-driven hydraulic governors were improved again in 1930 with Elmer's supervision of the design, development, and manufacture of the permanent magnet generator, which became the most popular type of governor head drive.[14]

Elmer Woodward was fortunate in the 1920s and afterwards to have a group of fine craftsmen working for the company. With his inventive skills and with the supporting ideas and excellent craftsmanship of George Sorensen, Bob Kelley, Tom Davis, Ward Ring, Jack Kurtz, Cy Erickson, and Dave and Fred Sandell, the company was the acknowledged leader in water wheel governing. Irl Martin during this time was struggling to learn the technology and the business and spent most of the 1920s working his way up from his position as an unskilled worker in the machine shop. Finally, after Elmer Woodward had asked Ward Ring to make Irl into a machinist, Irl was certified as a full-fledged (if modestly successful) machinist familiar with all the machines. Irl for his part found machinist work monotonous, but he recognized the importance of machining and developed an affinity for seeing the work from a machinist's point of view.

This attitude helped the company when he became general manager in 1929 because, whenever financially possible, the company replaced old, worn-out equipment with new top-of-the-line equipment. Irl, for example, had once worked on a lathe that "shook like a mangy dog" but still was serviceable to turn out parts good enough to maintain a tolerance of half a thousandth of an inch, the standard in the 1920s. But these tolerances were unsatisfactory to Irl. One day, not long after he had become general manager, he was watching a machinist named Art Fairgrieve operate a three-inch boring mill. The mill was running "pretty slowly" because Art explained, it could not be run faster since "the cutters wouldn't stand the speed." Irl then asked Art to buy some new, stronger cutters at Ingersoll. Without hesitation, Art stopped the mill and went to Ingersoll to pick up the cutters. Now the machine could be run faster, and in one day Art could machine as many parts with new cutters as the old ones could machine in two or three days.[15] This was a quick lesson in efficiency for Irl and, after this, machine upgrading began.

By the time Irl was made general manager, he had worked all the machines and had also been assigned in turn to engineering, accounting, road work, and sales. He was clearly the heir apparent, although Elmer's sons, Walter and Harold (Doc), were still working for the company during this time. It may have surprised some company members to see Irl promoted ahead of Elmer's sons, since one of them would seem to be the logical choice to succeed his father. But they were uninterested and apparently lacked the drive to run the company. Irl, however, was both interested and ambitious. He was also cocky and argumentative, and he and other company members

The gentleman in this 1925 photo of the Dixon, Illinois, I. N. U. Company hydro plant, is standing between a Woodward pump (left) and gate-shaft (oil pressure) governor.

often fell into heated debate over company policy. Some people in the company were put off by Irl's blunt outbursts and assertiveness, but Irl was never one to hold a grudge.[16] Moreover, he consistently looked for better ways of getting a job done while empathizing with the workers' concerns. In time, his attempts to make the company function like an extended family would pay off.

Although employees were increasingly treated as part of a family throughout the 1920s, the corporation remained closely held within the Woodward family proper. Only a few trusted employees besides the Woodwards held stock. The company had incorporated in 1902 with only three stockholders, and the only non-Woodward member had been quickly bought out. After Amos's death in 1919, and his wife Mary's death in 1921, the stock became more widely distributed within the family.[17]

The value of each share of this stock appreciated so rapidly in the 1910s and 1920s, that even to own a few shares was a lucrative investment. By early 1913, when capital and surplus totaled $136,000, each share had increased in value from the original $100 to $455. When Mary (Mate) Woodward died in 1921, the value had risen to $650 and, by the end of the decade, the value of each share was about $1,600.[18]

By 1928, only five people outside the Woodward family held stock in the company; Elmer himself owned almost half the shares (see Table Eight).[19] His own, and his children's shares, put a comfortable majority of all shares under his control. Elmer was "The Company" in most ways and was firmly in charge of what the company did. In running the company, he had the overwhelming support of all members of his family.

All shareholders were well-rewarded by the dividends paid out by the company during this successful decade. More governors were sold between 1920 and 1925 than in any previous five-year period. Moreover, because the newer hydraulic governors were more sophisticated, their higher prices increased the company's profit margins. By

TABLE EIGHT

1928 SHAREHOLDERS

Elmer E. Woodward	148 shares
Walter T. Woodward (son)	10 shares
Harold S. Woodward (son)	5 shares
Dorothy Woodward Martin (daughter)	10 shares
Frank M. Woodward (brother)	15 shares
Minnie W. Taylor (sister)	45 shares
Herbert M. Woodward (cousin)	10 shares
WOODWARD FAMILY	**243 SHARES**
R. J. Kelley (Vice President)	43 shares
Thomas Davis (Superintendent)	2 shares
Estelle Milne (Secretary)	2 shares
George Sorensen (Engineer)	5 shares
Cyrus Erickson (Foreman)	5 shares
COMPANY EMPLOYEES	**57 SHARES**

1923, the company was selling a wide variety of oil pressure governors (later referred to as gate shaft governors) as well as the compensating type mechanical governor. They also sold accessories like rotary pumping units, pressure control switches, motor-operated limit stops, pneumatic head level controls, float head level controls, automatic starting mechanisms, and motor-driven governor heads.

One popular type of oil pressure governor was designated as the HR (horizontal relay). It came in thirteen sizes that varied from the HR 2, the smallest, with 2,500 foot pounds capacity that cost $1,050 with a motor-driven pump, to the largest, the HR 60, with 60,000 foot pounds capacity that cost $4,900 with a motor-driven pump. There were also LR governors that came in eighteen sizes and two basic types. These varied in price from $700 to $3,000. Finally, the LHR governors came in twelve sizes varying in price from $825 to $3,950. Accessories cost more, although motor-driven governor heads were included free if customers bought the automatic starting mechanism that allowed the governor to operate in hydroelectric plants without constant attention. The most expensive governors did not have listed prices, but customers were encouraged to inquire. These were the type A actuator governor and the type G actuator governor. The former ranged in sizes all the way up to 300,000 foot-pounds capacity and was for use on large water turbines. By this time, the mid-1920s, some 5,500 Woodward governors were in use all over the world regulating an approximate total of 3,000,000 horsepower.[20]

Also during this period, Elmer's innovations were the subject of an article in *Electrical World* magazine. The article praised Elmer's past successes and current "invention, development and perfection of an apparatus required on governors for automatic, nonattending hydroelectric plants." The article went on to predict that Elmer's innovation would encourage the building of hydroelectric plants at water power sites that otherwise might not be developed since the new governors would not require workers to remain on site twenty-four hours a day. The *Rockford Register Gazette* in an excerpt from the article, gave credit to the "genius and success of this well-liked modest Rockfordite who has contributed so much" to Rockford's industrial prominence.[21]

For the rest of the decade, yearly sales were strong (see Table Nine).[22]

Early in the decade, almost all of the foreign sales were to Western Hemisphere countries—most to Latin America, and a significant number to Canada. There seems, however, no way to account for the tremendous drop in foreign sales in 1922, since overall sales were strong that year. In the next year, foreign sales rebounded strongly. In 1923, the largest unit sales year, sales to Canada were 65 percent of all non-U.S. sales. Canadians continued to buy well over 50 percent of the governors that Woodward sold to other countries for the remainder of the decade. In fact, in 1928 Canadians purchased forty-three of the forty-eight governors sold outside the United States.[23] Woodward's dependence on sales to Canada would continue into the 1930s but, once the full force of the worldwide Depression hit, foreign sales slowed markedly and practically stopped.

TABLE NINE

YEAR	TOTAL SALES	FOREIGN SALES	FOREIGN SALES AS % OF TOTAL SALES
1920	248	35	14%
1921	190	31	16%
1922	208	7	3%
1923	262	45	17%
1924	229	26	11%
1925	224	22	10%
1926	150	29	19%
1927	198	30	15%
1928	180	48	25%
1929	175	33	19%

The 1920s, in short, was the company's most profitable decade (see Table Ten) up to this point in time.[24]

Even more remarkable than these figures is a fact they do not reveal: the company's liabilities for each year were minuscule—just a few thousand dollars for accounts payable. Assets like real estate, buildings, machinery, and stock had already been paid for, and there was always a large surplus of accounts receivable, cash in bank, government securities, and the cash value of life insurance. The company was being run throughout the 1920s on a cash basis with no borrowing necessary, and this alone contributed to the generous dividends paid out during this decade (see Table Eleven).[25]

Benefits to the Woodward family were also evident. By standards of the 1920s, Elmer had become quite wealthy after nearly fifty years of hard work. As the largest shareholder, he reaped the greatest dividends. But even family members

TABLE TEN

DATE	NET WORTH	ANNUAL SALES
December 31, 1919	$188,205	$148,738
December 31, 1921	275,409	269,363
December 31, 1922	314,493	231,112
December 31, 1924	483,308	Not Available
December 31, 1926	489,883	Not Available
December 31, 1927	534,715	Not Available
December 31, 1928	484,475	Not Available
December 31, 1929	483,623	Not Available

who owned ten shares received $1,000 every time a 100 percent dividend was declared—a frequent occurrence in the 1920s. Robert J. Kelley, with ten shares in the company by 1924 and forty-three by 1928, also profited generously. Even Estelle Milne, who owned just two shares during the 1920s, got $200 every time a 100 percent dividend was proclaimed. With the 350 percent dividend in 1928, she collected an extra $700—undoubtedly a large boost in her total compensation for the year.

Another issue began to emerge in this decade that was to have an important influence on the company's future, and that was the matter of compensation for factory workers. Factory wages paid during this time were competitive with the Rockford average, and the closeness and camaraderie within the company promoted better working conditions than in most other plants. Despite this, the average employee at Woodward did not benefit from the company's prosperity in anywhere near the same proportion as Elmer, his family, and the worker stockholders. Irl was aware of this all during his first years with the company, and it was during this time that he probably began to formulate a more equitable way to include all employees in the profit distribution in a manner similar to the payment of extra compensation in 1919. No direct extra compensation was paid to the employees in the 1920s, however. In fact, wages were temporarily reduced in June 1922 due to lower sales and lower prices for governors.

TABLE ELEVEN

YEAR	DIVIDENDS AS A PERCENT OF $30,000 CAPITALIZATION	ACTUAL COST TO THE THE COMPANY
1920	35%	$ 10,500
1921	75%	22,500
1922	80%	24,000
1923	300%	90,000
1924	200%	60,000
1925	200%	60,000
1926	100%	30,000
1927	175%	52,500
1928	350%	105,000
1929	200%	60,000

The officers, likewise, reduced their own salaries as a matter of principle. Shortly afterward, the company did introduce two fringe benefits. In December 1922, Elmer, as president, and Robert Kelley, as secretary, who together made up the entire Board of Directors, decided to contract with The Travelers Insurance Company to ensure the lives of the officers and all the employees of the company. This company-paid insurance was in the amounts of $500 to $2,000, depending on how long the employee had continuously worked for the company. Five years later in 1927, the company began an insurance policy that paid accidental death, dismemberment, and disability income for all members.[26]

One more major innovation occurred during the 1920s—the introduction of governor schools in 1927. Like some other decisions

Participants in the Woodward Governor Company's first governor school, held in 1927. Woodward members included Elmer Woodward (No. 29), Irl Martin (No. 30), Roy Estell (No. 32), Dave Sandell (No. 33), and George Sorensen (No. 31).

made at the Woodward Governor Company, this one came about when an unforeseen event led to a good idea.

In the fall of 1926, Elmer and Irl took a pleasure trip by automobile through a portion of the Midwest. Since neither could resist stopping at hydro plants to see how the Woodward governors were working, it was more of a working vacation than actual time off from work. Their visits promoted good public relations, but what they found troubled them. In almost every case, operators praised the governors, but Elmer and Irl noticed that the governors were not properly adjusted to give the best results possible. This led to a dilemma. How could Woodward tell plant operators to make proper adjustments without hurting their pride or coming across as meddling outside experts who would show them how to do their jobs? Woodward's operating manuals were good and getting better, but some additional personal instruction clearly was needed. Now an opportunity had presented itself to provide this aid, and the governor school became the vehicle for the message.

Irl first had to convince Elmer that the school idea was a good one. Elmer was skeptical. It might be a fine idea, he thought, to invite the customers' hydro operators to Rockford, but he also believed that they were unlikely to make the trip. However, Irl countered, there was no other practical way to meet the mutual needs of the company and its customers, so Irl sent out invitations. His hopes were met when the response was "immediate and enthusiastic."

The first school met in February 1927 and, by everyone's standards, it was an immediate success despite one glitch. Irl, who was the teacher, began with the assumption that his students needed basic knowledge about how governors were put together and how they worked. His assumption was right, but then his presentation was interrupted by a long distance phone call, and Elmer filled in until he returned. Elmer, on the other hand, assuming that the students knew all about governors, proceeded to talk way over their heads. When Irl returned from his call and asked, "How far have you gotten?" the unanimous reply was, "Right where you left off."

Once customers got over their inhibition to criticize the governors made by their host, the first school became a practical learning experience for both students and teachers. It also led to better governors and better relations with the companies and their representatives who purchased the governors. The exchange of information was the key. In retrospect Irl said, "Many of the improvements in governor operation and wear, as well as new developments, have resulted from an analysis of troubles brought up for discussion" at the governor schools.[27] Since 1927, the schools have been a company institution and are held periodically in Woodward Governor Company plants throughout the world.

A letter of appreciation in March 1928 described the school that year as "a howling success" and noted that the social benefits of the visit were not inconsiderable either. "What I am getting at is this—the big benefit that I received from the school was in meeting the Woodward Co. and their shopmen, and most especially the sixty men who were there, each one of whom had waterpower and electrics up against his heart."[28] All the advertising in the world could not have matched this sentiment nor provided as many potential advantages to the company.

The military wreckage of World War I was followed a decade later by the economic wreckage caused by the Great Depression. Few people were prescient enough to foresee the coming economic disaster, much less deal with its causes.

Banquet for the third governor school, 1929, at the Nelson Hotel. Elmer Woodward and Irl Martin are seated in the left foreground. All members of the company, including the janitor and truck driver, were invited to the banquet by Elmer.

When Herbert Hoover was elected president in 1928, he promised "a chicken in every pot, and two cars in every garage." All but the most wary believed that prosperity would continue during his administration. Certainly the leaders of the Woodward Governor Company looked to the future with confidence.

Confidence, in fact, was the recurring theme in the company's promotional magazine, *Water Power Control*, which continued to be published sporadically during the late 1920s and early 1930s. Irl edited it, and its contents were always optimistic. In April 1928, he used the Rotary Club motto, "He Profits Most Who Serves Best," as an example of what the Woodward Governor Company always sought to do. He especially encouraged customers to purchase the knowledge and experience of the company, and not just its machines. All they had to do to get this knowledge was to ask for it.[29]

By mid-1929, *Water Power Control* was promoting two new Woodward developments—a mechanical generator brake valve and a limit switch. The same issue made lighthearted comments about that year's governor school, but also noted that "despite the horse play...this conference was of a serious and instructive nature" and was especially useful in disseminating knowledge about governing principles that aided both the users and the manufacturer of the governors.[30] The big news in sales that year was the purchase of four Woodward type A actuators by the Amtorg Trading Corporation of New York. The news was big because the Soviet government owned Amtorg, which traded freely in the United States even though the United States government did not diplomatically recognize Joseph Stalin's regime. Woodward was pleased that its good reputation had spread to the Soviet Union and that its product would be installed on the 84,000 horsepower hydroelectric turbines of the Dnieper River project, the largest power plant in the world.[31] In fact, based on total horsepower, the Soviet purchases in 1929 made that country the company's third most important customer after Canada and the United States.

Rather appropriately the company ended the 1920s after being the subject of a *Rockford Register Gazette* promotional piece honoring sixty years of governing. It noted, "White coal is the magic name given to the waters of the great streams which have been harnessed and made to develop millions of horsepower." The implication was clear; the Woodward Governor Company was a prestigious company doing important things for the country while keeping ahead of the technical changes then going on in the industry. At that point, no one knew how much the company would be forced to change its major products when the market for them declined dramatically during the Depression years.

From Depression to Expansion

1930-1939

Necessity became the mother of invention for the company in the early 1930s. Without the new products developed during those years, it is very likely that the company would have joined the long march of business failures during the Great Depression. But, after some lean years in the early 1930s, new products would all but end the Depression for the company by 1935 and, in the years that followed, the company would experience unprecedented prosperity and expansion. On the eve of World War II, the company would also begin the transformation that would make it unique among corporations in the United States. However, just as the changes inspired by Irl Martin began to take hold, the company lost the bulwark of its seventy-year existence when Elmer Woodward died at age seventy-eight, on December 31, 1940.

Prosperity in the America of the 1920s ended abruptly in the decade's final months. Leaders lost confidence and business contracted after the stock market crash of October 1929. Although President Hoover actively tried to end the Depression, the country was overwhelmed by the rapidly-deteriorating economic situation. The federal government was reluctant to involve itself in economic events. This attitude reflected the conventional wisdom that individuals and private businesses were responsible for their own successes or failures.

By 1932, the Depression's cruelest year, the economy had bottomed out and the Hoover administration was floundering. Stock prices on the New York Exchange averaged 11 percent of their pre-crash

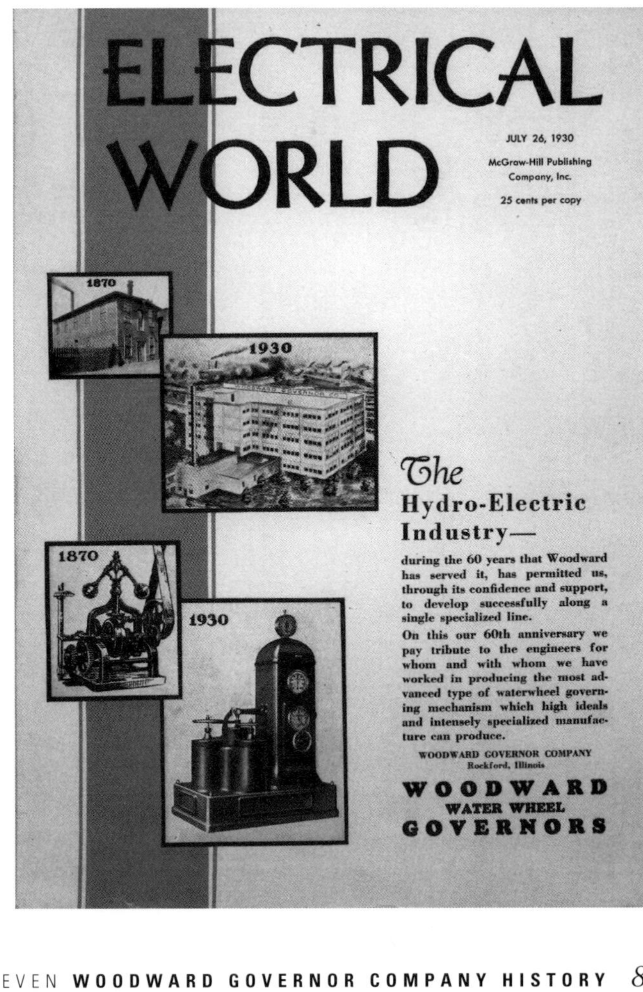

Woodward Governor Company advertisement in Electrical World *magazine, 1930.*

value. Investors had piled up paper losses of $74 billion between 1929 and 1932—a figure three times the entire cost of World War I. Over 5,000 banks had failed (including some in Rockford), and 86,000 companies had gone out of business. Unemployment was at an all-time high (over 12 million people—nearly 25 percent—were unemployed) and, even for those who had work, wages averaged only $16.21 per week.[1] Businessmen, heroes of the 1920s, were now scorned and abused. Hoover was no match for New York Governor Franklin D. Roosevelt in the election of 1932.

Eight years later much had changed, largely due to America's capacity to weather economic hard times. Roosevelt was so popular that he broke the unwritten two-term limitation for presidents and won re-election for a third term in 1940. However, for all his successes and his personal popularity, he only alleviated the Depression. The New Deal and its alphabetical agencies had often made things better, but at the cost of making the national government, in the eyes of many observers, too large and too powerful. The Depression itself continued to the end of the decade, after which government spending during World War II slowly pulled the country out of the decade-long slump and brought renewed prosperity. People who lived through the 1930s could never forget the anguish and frustration they felt, and few of them were neutral about the country's leadership. The view that FDR had saved capitalism was opposed by the view that the New Deal had transformed America into thinly disguised socialism.

Rockford's problems during the Depression mirrored the country's, much to the chagrin of those who had believed that Rockford was "depression proof." The third largest city in Illinois, which in the 1920s was described as "the most prosperous city in the nation," was now considered to be the "hardest hit of all" by the Depression.[2] Both descriptions were exaggerations, but the city was struggling during the early 1930s. A prime example was the decline in residential building permits from the 1920s to 1934. From 1922 to 1930, the value of new residences each year had been over $1,000,000. Then, in 1930, the total for 158 new apartments and private homes dropped to $662,000—less than half the 1929 figures of 368 buildings valued at $1,587,000. The drop steepened in 1932, when only 11 permits valued at $37,300 were issued. In the next two years the decline continued until a low point was reached in 1934, when only two new residential buildings were constructed.[3]

During these years, industrial activity also declined precipitously; those people still working were forced to take regular salary or wage cuts. Although this was matched by a drop in prices, to those without money prices were still too high and thus consumer spending plunged. In mid-1931, Rockford's civic leaders sought to end the local depression by staging a jubilee week extravaganza to encourage citizens to buy local goods. Bands were assembled, and a parade was planned for June 15 to celebrate the victory. But the celebration was premature, for that very morning three of Rockford's eight banks went bankrupt. The parade still went on before a good crowd, but it was hard to separate the spectators from the people waiting in vain to get funds out of the closed banks. One bandleader with a realistic sense of humor had his band play "I Can't Give You Anything But Love, Baby" as it passed by one of the closed banks.[4]

Until the New Deal brought in federal aid, beginning in 1933, Rockford was dependent on local and state relief. These were inadequate, and public welfare funds ran out in 1932. In 1933, Rockford was aided by the federal Civil Works Administration; in 1934, by the Federal Emergency Relief Act; and in 1935 and afterwards, by the Works Progress Administration. Thousands of people continued to get various forms of federal government relief in 1940. By this time, due to national defense needs and the new selective service law, Camp Grant was being rebuilt—a project that required 6,000 workers and cost over $8 million. Just as in World War I, this helped reinvigorate Rockford.[5] Also, during this time, hundreds of homes were built and businesses began to expand. Woodward was a part of this trend.

Although the company was strong in 1930, it was already beginning to feel the effects of the deepening Depression. Its economic problems worsened in 1931, although it still remained in the black. However, in 1932—the worst year of the decade-long Depression—Woodward lost money, and losses continued for the next two years. Only after its new products for diesel and aircraft engine governing came on line did the company return to the black in 1935. Then, despite the continuing Depression, sales and profits began to skyrocket. From then on, the company left the Depression far behind. This led to a large influx of new staff, an expansion of the Mill Street plant, and ultimately the decision to build a new plant after outgrowing the old one. The company's successes by 1935 were astounding when set against the real financial danger of the 1931-1934 period. Table Twelve vividly shows how close disaster came before new products began to transform the company.[6]

At the beginning of the decade, no one knew how long the Depression would last, but most people remained optimistic that it would be short-lived. Certainly Woodward Governor Company leaders betrayed no great sense of alarm because orders and profits remained satisfactory throughout 1930, even though both declined compared to 1929. Nevertheless, the company remained busy and even spent a great deal of time in early 1930 revamping the factory for a more efficient operation.[7]

The company publication, *Water Power Control,* begun in March 1928, continued sporadic publication. Irl Martin was its first editor,

TABLE TWELVE

YEAR	GROSS SALES	NET INCOME AFTER TAXES (LOSS)
1928	$ 335,000	$ 66,000
1929	318,000	65,000
1930	271,000	39,000
1931	329,000	31,000
1932	139,000	(17,000)
1933	64,000	(16,000)
1934	128,000	(4,000)
1935	289,000	38,000
1936	526,000	97,000
1937	933,000	170,000
1938	803,000	153,000
1939	1,396,000	308,000
1940*	1,800,000	328,000

** Woodward changed to a fiscal year ending September 30 in 1940, so these are yearly figures through 1939 and nine-month figures for 1940 only.*

and much of its content dealt with past and future governor schools. Irl related anecdotes about people who attended them, and threw in a modicum of technical information about better governing. The governor schools (called Water Power Control Conferences from 1929 to 1931), proved to be both a practical and a goodwill bonanza for the company. The Woodward Governor Company always promised customer satisfaction; the conferences enabled the company's staff to get to know their customers and potential customers in an informal way. In 1930, an impromptu baseball game was umpired by Monsieur Jardy from France, who spoke no English. Irl reported that the results of his first decision reminded Jardy of "the toughest days of the war." After the supposed removal of the Canadian contingent for playing rugby, the Canadians redeemed themselves on the sideline by producing individual portions of "strange smelling tea" which broke up the ball game. The camaraderie was distinctly apparent in this boy's club atmosphere as Rockford was invaded by a regularly increasing number of individuals representing utility companies, consulting engineers, electrical manufacturers, water wheel manufacturers and even higher education institutions from the United States, Canada, and Europe.[8]

Woodward's reputation as a company that cared about its customers was not earned only in Rockford, however. Shrewd salesmanship, boldness and hustle, and a close attention to detail also contributed, as the story of the Saluda

Development project shows. The project was started in 1928 by the Lexington Water Power Company of Columbia, South Carolina. This power project, situated at the world's largest earth-filled dam for hydroelectric purposes, eventually bought Woodward type A actuators, pilot exciter drives for governor head motors, and a semi-automatic operation with "dual control." Before it made these purchases, however, it had first chosen Woodward competitors—but that was before it had heard Irl Martin's pitch on dual control. The story of the development of the dual control combines the guile, brashness, and confidence of Irl Martin with the ingenuity and pragmatic adaptability of Elmer Woodward, for no such thing as dual control existed anywhere in the industry.

Irl had gone to the engineering office of the Lexington Water Power Company to bid on the portion of the job that required a governor, but found himself face-to-face with representatives of three water wheel companies, Allis Chalmers, I. P. Morris, and S. Morgan Smith. They pointedly told him that Woodward had no chance for meeting the project's governor requirements because they were going to design a governor among themselves and get not just the water wheel order, but the governor job as well. After waiting all day to see the chief engineer, Irl was told that the contract for the governor had been let. Nevertheless, the chief engineer agreed to reward Irl's patience by listening to his sales pitch.

While engineering was not Irl's strong point, he was both a good listener and a good salesman. He heard the engineer's complaint about the difficulty of controlling the huge power units between the generator man on the floor and the switchboard operator some distance away. Ordinarily they communicated by phone, and any disagreement between them delayed getting a unit on line. Irl's solution was dual control, in which the switchboard man had complete control unless the generator man activated a safety shutdown switch and took over in case of an emergency. The engineer, like Irl, had never heard of dual control, but exclaimed, "That's exactly what we want; your competitors never mentioned that method." Irl blithely explained that they could not possibly know of the mechanism, since they were water wheel specialists not governor specialists like Woodward. The engineer then switched gears, promised the job to Irl and called in the angry water wheel representatives. Irl quoted a premium dual control price to the water wheel representatives. Since they wanted the water wheel part of the project, they swallowed hard and agreed to his price.

When Irl returned to Rockford, Elmer congratulated him on his success, but was dumbfounded when Irl said, "In order to get the job, I had to sell them a dual control." Elmer replied, "What is that?" After being told the hazy details and Irl's tentative plans on how to accomplish them, Elmer said, "You have got to quit selling things that you don't know anything about." Irl replied with confidence, "Pops, you can make anything, and I can sell it." Within a few days, Elmer was using his small broken ruler to make measurements, and Irl knew that they "were in business." Dual controls soon became standard, and the company furnished them on numerous installations.[9]

By the time the Saluda project was completed Elmer had shown his confidence in Irl in a number of ways, not the least of which was to loan him money. In fact, on May 31, 1930, Irl was allowed to buy twenty shares of Woodward stock from Elmer for $24,000. The stock was held as collateral by Elmer until Irl paid the note plus interest at 6 percent in November 1937.[10] The price in 1930 undoubtedly was less than the stock's worth, but Irl must have had a few second thoughts as the company teetered on the edge of the abyss in the next few years. However, since Irl was general manager, it also made sense for him to have a stake in the company's future. In 1932 he was elected second vice-president and, in 1934, he was a company director, secretary, and treasurer.[11] Thus, in the early 1930s, Irl was becoming more and more responsible for the company's welfare.

Perhaps the best example of the synergistic relationship that had developed between Elmer and Irl occurred after the bank crisis in Rockford in June 1931, when three banks failed. The outcome of this catastrophe has been the subject of a growing company legend that highlights the selflessness of the management and all of the individuals it employed.

Among the failures was the Rockford National Bank, the depository for about $15,000 of the company's money as well as the personal accounts of many of the company's workers.[12] When this occurred Elmer was not in Rockford and, in his absence, Irl called a plant meeting to discuss what should be done. Management's first decision was to make available the petty cash drawer for anyone who needed a small amount of instant cash. However, the big problem was to preserve everyone's job at a time when the company's orders were rapidly declining and few new orders appeared to be likely in the foreseeable future. The solution turned out to be deceptively simple, and derived from the company's long-standing commitment to humane business values.

The normal work week at this time was still sixty hours, and much of the work was not scheduled for shipment for months, due to the long lead time it took to bring all contractors' products together for one project. The dilemma faced Woodward management with a difficult choice between two equally unattractive alternatives. One, either to complete all work as soon as possible, risking a company layoff if no new orders came in; or, two, work fewer hours but make the work last longer and guarantee employment for a longer time. Someone suggested that working hours should be reduced to twenty per week. The group unanimously decided that the second choice was the proper solution, even though their income would drop by about two-thirds.[13]

When Elmer returned, he endorsed their decision and made a further personal promise: as long as either the company or he were solvent, he would guarantee the individual incomes at the twenty hours per week level. This was a magnanimous decision, one that could have wiped him out financially. In these pre-New Deal days of 1931, there were no federal programs to aid the unemployed, and the state and local welfare organizations were vastly underfunded. Elmer, in effect, became the equivalent of a one-man welfare program regardless of orders, so long as people put in their twenty hours. His humanitarian gesture was reciprocated in kind by the company's membership. They usually worked much longer than the required twenty hours, and often used their extra time on the job to clean the machines and paint the inside of the plant. Out of near financial disaster, then, came the beginning of Woodward's pride—an immaculate machine shop.[14]

Orders began to pick up in 1933 and 1934, and the shop was ready to go back to sixty hours per week. But, by this time, much of Franklin Roosevelt's New Deal legislation had passed, with significant effects on the ways business could operate. The law that most affected Woodward was the National Recovery Act. In its attempt to provide jobs, the law made forty hours the maximum legal workweek. Woodward, however, had enough orders for sixty-hour weeks and, because of the long time it took to train machinists, new employees could not be hired to pick up the hours lost to the established factory hands. The machinists at Woodward solved the problem by punching out after forty hours and then returning to work on their own time, and without compensation. Perhaps they bent the law, but this was their way of paying Elmer back for his generosity when times were tough. This decision reciprocated the solidarity that Elmer had previously promoted, and helped lead to a special and unusual management-worker relationship at the company which has endured both in legend and in reality.

Looking back to this time, Irl often exaggerated the repercussions of Elmer's decision to guarantee incomes by claiming that it put Elmer and the company "heavily in debt for the first time in many years." Irl also claimed that Elmer drew "no salary during the entire period of the Depression." Elmer, it is true, did use his resources for loans and declared a moratorium on home mortgages that he had financed out of his own money. He cancelled all payments and interest until the shop normalized hours and pay. When taxes and winter fuel costs became due, both were paid for out of company funds. He also borrowed money himself to aid his workers. But, in fact, neither he nor the company went heavily into debt. Both his finances and the company's remained solidly in the black despite company losses in 1932,

1933, and 1934. Elmer also drew a salary throughout the 1930s, although it regularly declined from its 1929 high of $21,000 to $9,000 in 1934. It began to climb only after the company returned to prosperity, reaching $24,000 in 1937. Irl may simply have confused salary with company dividends, since the latter were eliminated from 1932 to 1936—the only decision both fair and responsible considering the sacrifices of all concerned.[15]

The company not only made it through the tough early years of the Depression, but it thrived once Elmer and the other engineers, Al Kalin, Walt Thorell, Bill Whitehead, Norm Noling, and George Sorensen, began to see new applications for governors that previously had been used only for hydro projects. Even before these changes, Kalin made his value to the company felt with the permanent magnet generator governor drive. This drive superseded all previous drive methods, such as the pilot exciter drive, and it was first installed commercially at the Alexander Power Development on the Nipigon River in Ontario, Canada, in 1932. Unlike the pilot exciter drives, it caused "the governor motor to follow faithfully a rapid acceleration of the main unit and thereby the governor is up to speed the instant the turbine is up to speed; there is no loss of time and no overspeeding." This invention became standard equipment on all large turbines.[16]

The permanent magnet generator success touched off a frenzy of activity at the company in late 1932 and early 1933. Sales were dropping rapidly, the company was losing money, and few hydro projects were planned in the foreseeable future because nearly all of the good rivers were already in use. The future looked bleak for the company unless it could diversify into governors for other prime movers. Luck and inadvertence, not to mention shrewd marketing and excellent engineering, combined in 1933 to lead Woodward to build governors for diesels and for aircraft engines. And, once those opportunities were open, brilliant engineering led by Elmer Woodward brought forth the new products.

Company confidence was increasing by the time the sixth governor school—now renamed Prime Mover Control Conference—met in mid-1933. For the first time, the conference was divided into three sections—hydro, steam, and diesel—followed by a summary session held jointly on the coordination and control of system prime movers. The session on steam represented another area of diversification for Woodward; early in 1933 the company had made some important breakthroughs in diesel and steam governing.[17] The most important, though, was still in gestation, and that was Elmer's perfection of a governor to control the pitch of the blades of airplane propellers.

The diesel governor was first, and it came about, not through company research, but because of an unusual governing problem at an installation in

Canada owned by Fairbanks-Morse, a diesel engine company with a plant in Beloit, Wisconsin, just twenty miles north of Rockford. Fairbanks-Morse had written Woodward that their Canadian plant's water wheel, equipped with a Woodward HR governor, was running in parallel with one of their diesel engines. Theoretically, the diesel engine, being faster, should have been doing the regulating, but the Woodward governor was so sensitive that it took the regulation away from the diesel. What could be done?

To investigate the problem, Elmer got permission to use the Fairbanks-Morse facilities in Beloit and brought with him a control standard and cylinder from a 2,500 foot-pound water wheel governor. He mounted these on the end of a type Y Fairbanks-Morse diesel engine. The governor Elmer built was strictly a prototype but, through experimental trial and error, he was able to get the engine and governor to work together, while leaving the regulating functions to the engine. From this prototype, Elmer developed the Woodward IC (internal combustion) governor.[18] The company eventually sold thousands of them for stationary diesel electric units. Appropriately, Fairbanks-Morse bought the first ones in late 1933.

Woodward expected that its IC governor would be made in one size for central stations, but quickly realized it would have to be adapted to other types and makes of diesel engines. In fact, the company soon developed a much smaller governor, the SI, for railroad and steamship service. SI stood for steam and internal combustion, and it was developed after the Winton Engine Company of Cleveland, Ohio, inquired whether Woodward could build governors for engines Winton was developing for railroad service. (The Winton Company later became the Electromotive Division of General Motors.) The Winton episode illustrates again typical Woodward ingenuity and engineering skill, and parallels the occurrences that took place with the Saluda Development project.

Elmer Woodward installing the first trial governor on a diesel engine at Fairbanks-Morse Company, Beloit, Wisconsin, in 1933.

Woodward was once again asked by a potential customer to do something it had not considered

Elmer Woodward, Roy Estell, Irl Martin, Al Kalin, and W. E. Brill (from Winton Engine, a Cleveland diesel engine company) at the back door of Elmer's house at 817 Haskell Avenue.

doing on its own. However, once the need was seen, the ingenuity of Elmer and the other engineers quickly satisfied the customer's needs. Irl Martin served as the sales engineer on the Winton project. As he left for Cleveland, fully expecting to sell an IC governor, he was admonished by Elmer, "Don't forget that if the type IC governor won't handle the job, we don't want the business." This was a limitation that Irl was not prepared to accept, despite his affection for Elmer; the company definitely needed more business.

At Winton, Irl met with the general manager, and it quickly became clear that the company's needs could not be met with an IC governor, which was much too large. Moreover, while the general manager was enthusiastic about Woodward's interest in building a new governor, his chief engineer resented an outsider coming in to do something he felt Winton could do by itself. He sarcastically informed Irl that Winton could make as good a governor as Woodward. Irl mulled this over and decided to call the engineer's bluff: "Fine," he said, because "Elmer didn't want to sell Winton a new governor anyway." The engineer did some mulling over himself, and then rewarded Irl with a quick retreat and the reply Irl wanted to hear: "Listen, ... I'll tell you who's going to build the governor for us and you're going to do it."

Now that Winton, like Saluda, was talked into buying a governor that did not exist, Irl's next problem was to sell the idea to Elmer. Elmer could be stubborn, like the Rock of Gibraltar, but Irl argued that refusing the project would tarnish the company's "reputation for integrity and capability." Both of these were dear to Elmer's heart, because both his and the company's reputations were based on them. Therefore, he gave his assent, and the project which resulted in the SI engine governor was begun. The first of these was completed later in 1933 and, with varying modifications, it continued to be sold for over two decades.[19]

One of the earliest applications of the SI, was on yachts. The governor's main requirement was to reduce vibration, especially on twin screw vessels. This required that an SI governor, adapted for marine use, be mounted on each engine, with the engines connected by electrical equipment. Al

Kalin was especially prominent in working on this problem, and he developed a method of synchronizing the rotation of the two engines' shafts. This was called "synchrophasing," for which he received two patents. The governor kept the tips of the blades of the propellers in synchronization so that they "would pass the hull in an opposite direction of rotation at exactly the same instant." Synchrophasing not only reduced vibration but, more importantly, Kalin did it with an inexpensive governor rather than with expensive electrical equipment designed especially for the purpose.[20]

This invention was installed on the diesel yacht "All Alone" in 1933, and on the "Chalena" in 1934. The results were impressive—one observer described the Woodward control system as "almost uncanny in its operation" and noted "whilst I have always had a deep respect for your equipment and your method of manufacturing it, I have even a deeper respect for your inventive ingenuity. Congratulations." Daniel Cox, of Cox and Stevens, a firm of yacht architects, was especially enthusiastic and endorsed the product saying, "Fortunately, the cost of your installation is by no means great and as the reduction in vibration will add enormously to the value and comfort of any yacht, we believe there is a very large field open for your synchrophasing installation."[21] In the future, this invention would be used on a large number of private yachts, but it was also sold to the U.S. Navy in large numbers by 1934 and 1935. The SI governor and its adaptations became so popular, that Elmer once reported, with tongue in cheek, "Enclosed is the order for eight SI governors. The order was forced on me. I must, therefore, decline to accept the promotion although I am honored by the offer."[22]

Elmer Woodward (far right) with Hamilton Standard engineers. Elmer had taken his experimental propeller governor model to East Hartford, Connecticut, for installation on an aircraft.

As the company was gearing up to sell its new IC and SI governors in 1933 and 1934, the Hamilton Standard Division of United Aircraft contacted Woodward to help them make their adjustable pitch propeller automatically adjustable through the use of a governor. Simply to note that the effort began on March 25, 1933, and was successfully completed on March 26, 1934, trivializes the inherent difficulties of this project. This was also new territory for the company and, to complicate matters, the company and its engineers were preoccupied with the new diesel governors. At first, they tried to adapt the SI compensating governor to the airplane engine. This proved too slow, too large, and too expensive for Hamilton Standard's needs. Just as Woodward faced failure, Elmer got involved and suggested that a non-compensating governor (like the original 1870 patented governor) would be appropriate. This would solve the weight problem, but since none of the engineers

George E. "Bud" Sorensen and Elmer Woodward inspecting airplane propellor governors in 1937.

were familiar with this "old-fashioned" governor, Elmer undertook the project himself. Meanwhile, the company's machine space was almost exclusively reserved for making the ICs and the SIs, and there was no room available for development work on airplane governors. Elmer solved this problem in a way uncharacteristic for the major stockholder and boss. He contacted an independent tool and die shop that was operated and owned by a former Woodward superintendent and subcontracted the experimental work to him. Elmer then spent much of the winter of 1933-34 walking back and forth to this shop with his pockets so full of castings for the small propeller governor, that he was unable to button his coat despite the occasionally severe weather. The project had become his responsibility, and it had probably never occurred to him that someone else ought to have done the leg work (Elmer was nearly seventy now). His policy had always been, "old customers first, new customers second, prospective customers last" and, therefore, the decision was easily made.[23]

Once the propeller governor was ready, he left for Connecticut to demonstrate it to Hamilton Standard. While Elmer was confident, Irl and some of the engineers believed that he would have his first major failure at an age when most people had already retired and were forgotten. Instead his tiny three-pound governor was an unqualified success. In the next twenty-five years, the company built over 400,000 of them, under an exclusive contract with Hamilton Standard that was the essence of simplicity, because Elmer was so trusting and fair in his business practices.[24] The contract consisted of three statements: Woodward would supply Hamilton with all of its propeller governors. Hamilton would buy all of its propeller governors from Woodward. Woodward would not sell propeller governors to any other customer.

With these new governors firmly entrenched in the market-place, Woodward was no longer dependent on making governors for water wheels. The new governors led to rapid expansion in personnel and an enormous increase in net income in the mid-to-late 1930s. Personnel rose from 32 in 1933 to 38 in 1934, doubled to 75 by late 1935, more than doubled again to about 180 in 1938, and by March 31, 1940, topped out at over 300. In seven years, through the worst depression in history, this was a staggering increase of almost 1,000 percent. Profitability was no less spectacular: after the losses from 1932 to 1934, net profits increased from $38,000 in 1935 to over $300,000 in 1939. In fact, the

Individuals shown in front of the Bonneville Dam governing unit in 1937 are Woodward members on left of platform: Bill Taylor, Irl Martin, Norm Noling, and Al Kalin; six visitors from Edison Electric Institute join Elmer Woodward (far right on platform); and more Woodward members on the right side of the platform: Bob Kelley, Luke Avery, Walt Thorell, ----- Larson, and Ed Lilja. Because of the large size of the Bonneville cabinet actuator, it was assembled at Ingersoll Milling Machine Company.

$170,000 profit in 1937 was by far the largest in the company's history and, even after that, records were broken again in 1939 and 1940 (see Table Twelve).

The new products also required the company to change the name of its sporadically published magazine (still under the editorship of Irl) from *Water Power Control* to *Prime Mover Control* with the April 1935 issue. After June 1935, no issues were published until January 1937 when Norm Noling was drafted to take over as editor. Irl's participation continued with a last page column of news and gossip he called "The Last Word."[25]

Although governors for water wheels had become less important to the company by the mid-1930s, they remained a staple product and, in fact, the company produced at least one modified governor breakthrough in 1935 with the cabinet actuator. Production of the WO diesel governor was also begun that year, but the WO was not as successful as the IC and SI governors.

The cabinet actuators, however, were used extensively in the large hydro projects that the federal government began in the mid-to-late 1930s, most notably at Hoover Dam and Bonneville Dam. They were designed to consolidate the principal components of a governor into a "neat, compact and easily operated unit," and Woodward declared that they were the company's "most important contribution to the art" of hydro regulation in many years. The units were attractive in their self-enclosed cabinet, easy and efficient to operate and service, and came with a sophisticated

Elmer Woodward, in 1937, compares the smallest and largest Woodward governors: the airplane propeller governor in his hand and the Bonneville Dam cabinet actuator.

array of auxiliary controls. And they sold well: twenty-three were installed between 1935 and January 1937.[26]

The Paint Department, Mill Street plant, in 1935.

The Assembly Department, Mill Street plant, in 1935.

Even as the company expanded to accommodate orders for hydro, diesel, and aircraft governors, it remained a close knit and personable company. Its family environment, in fact, revelled in some curious and often funny antics. Expansion had lowered the average employee age, which raised the imaginative and good-spirited high jinks that naturally come with youth. One popular pastime was to straighten copper tubing with a drill rod and hook it up with compressed air to make a crude gun. The pranksters would add sights and amuse themselves by shooting at birds. Sometimes amusement crossed the threshold to excess. One sharpshooter missed his feathered quarry, and the missile arced across Main Street and broke a large plate glass window at the International Harvester store. This caused a great flap, but the antics continued. On another occasion, Urban Floor, a young engineer, was struck on the neck. He reached back, felt the horrifying evidence, and concluded, as anyone stung by similar proof would, that he had been shot. And he had, by a non-lethal grape. One day someone put a whistle bomb in a test stand. It went off and caused panic among those not in the joke, since it acted like an electrical fire. The most popular on-site recreation was a crude and weighty variation of bowling, using large forty- to fifty-pound cast iron counterweights as bowling balls. A good throw would take a ball the length of the third floor. No mention is made of what the bowlers used for pins, but no amount of creativity or sheer, unmitigated boldness could have topped the choice of Elmer Woodward himself, who came through the door just as a ball was bearing down on him. Without a word, he hopped up, cleared the ball, and kept on walking, much to the relief of the perpetrators who expected, at the very least, a mild, fatherly reprimand that could be heard all over Rockford. But chewing outs were not in Elmer's nature. In

Engineers and draftsmen on the fourth floor of the Mill Street plant, in 1936.

The Machine Shop, Mill Street plant, in 1935.

fact, perhaps, the only time he did swear occurred when an oil pump's valve was inadvertently left open. When Elmer turned it on, it erupted. The oil hit the ceiling and descended like an umbrella, inundating Elmer. After Art Nielsen turned it off, the unflappable Elmer commented, "Damn—I guess I'll have to go home and change clothes."[27]

This response was much in character for Elmer, whose even-tempered ways contrasted sharply with Irl Martin's fiery, occasionally explosive, personality. He was a man with a temper, the stomach to use it, and the skill to know when. But Irl had his soft side as well; he was particularly solicitous and protective toward the old-timers in the company. They had been around for decades and they had suffered with the company through the crisis between 1931 and 1933 by putting in a great deal of free time for the company. Now, in their advanced years, some of them could no longer do their jobs as well as younger workers. Irl always protected their jobs and their dignity.[28]

Elmer was the oldest "old timer," having started at Woodward when he was fourteen in 1877. He was the most dignified of all but, on one occasion, not even Irl could protect his dignity. He and Irl were on a business trip to New York and, for the trip between Buffalo and New York, they had rented a bedroom compartment on the train. Irl sat in the smoker while Elmer prepared for bed but, on returning to the room, he discovered Elmer was gone. Shortly thereafter Elmer emerged, "his knees rather wet and dirty and in a rather disheveled condition," from beneath the stationary train. He had been cleaning his "store teeth," he explained, and had left them on the lavatory when he folded it back into the wall. Not knowing that they fell into a receptacle provided for such forgetfulness, he assumed they had passed through the drain and were on the tracks. "Being very anxious to save his teeth, he ran outside and down on the ground next to the car and started hunting for his teeth." Irl was appalled because the train could have moved, but he attributed the whole episode to Elmer's age: "He is normally so slow moving and cool that it was quite a surprise to me that he would do a trick of that kind..."[29] It was certainly true that Elmer, as he reached his seventies, had slowed down physically, but his mental faculties, as proven by his inventions, continued as strong as ever. The fruits of his lifetime of labor in the narrowly-chosen field of governing led to a plentiful harvest of gratitude in the last few years of his life. The accolades began on October 9, 1937, when Elmer was honored on his seventy-fifth birthday for his sixty years of service to the company his father had begun. The celebration was held at Rockford Country Club and included all of the 160-person work force plus about forty local and regional businessmen and guests with ties to the company. Even Elmer, who by nature had long sought to avoid personal publicity and public ostentation, was "visibly pleased" by this display of gratitude and friendship. Elmer was presented with a silver and mahogany plaque with each member's signature etched into the metal. The inscription summed up the consensus on Elmer's character and influence:

For the Wisdom, Patience, Sincerity, and Humility with which you have fathered our mutual welfare through the years, we dedicate this plaque as an expression of affection and appreciation which words cannot convey.[30]

Honoring Elmer's birthday became a yearly tradition at Woodward after 1937. For many years, the fall profit sharing party was held on October 9.

A few months after his 1937 honor, Elmer spoke at the national spring meeting of the American Society of Mechanical Engineers in Los Angeles, California, on modern turbine governing. His topic emphasized the great advance in water turbine governing brought about by the cabinet actuator, which "combined as much as possible of the governing system into one unit instead of scattering it around the plant."[31] By late 1938, Elmer was honored again when he was made an honorary member of the Rock River Valley Section of the American Society of Mechanical Engineers (ASME).[32] This followed Elmer's birthday celebration by two months. For that celebration, Carl Chandler, a company machinist, wrote an affectionate poem that emphasized Elmer's impact on all who worked with him.[33]

Elmer Woodward's 75th birthday celebration at Rockford Country Club, October 9, 1937. Shown here (left to right) are: George Sorensen, Dave Sandell, Estelle Milne, Elmer Woodward, Cy Erickson, Bob Kelley, Ward Ring, and Fred Sandell.

There's a man at our shop, who lives in my mind,
He's generous, patient, considerate and kind.
He's one in a million, I don't have to say,
That he made this shop what it is today.

He never would take, more than he'd give,
And was always content to live and let live.
He earned what he liked in his struggle for pelf,
He made others happy as well as himself.

There were times when he worried a little, I guess!
But he hit that old ball and kept doing his best.
I'd like to be like him and carry that smile,
Just to pattern his life and do something worth while.

He's probably human, and not without fault,
But he's toiled on this earth and been part of its salt.
He lived a clean life and never did shirk,
Just an Abraham Lincoln at the shop where I work.

A few years back at the sound of the gong,
They were all good workmen, but just seventeen strong.
He watched over these fellows and kept them in sight,
Their work was his work, their fight was his fight.

He's alert and still going at seventy-six,
Still striving to learn and knows most of the tricks.
He's a very fine person as I've said before,
God help him to live to a hundred or more.

I can't buy a present, at least not today,
But many happy returns in my own simple way.
So, if you guess who I'm praising and think he's a man,
Just get up on your toes and let's give him a hand.

by
Carl W. Chandler

Presented to Mr. Elmer E. Woodward on his 76th birthday, by employees of Woodward Governor Co., Rockford, Ill.

In early 1939, the honors continued when Elmer was made a fellow of ASME, the highest award presented by the society. The national president of ASME made the presentation and declared it was for Elmer's "unexcelled work in the fields of mechanical engineering."[34]

Elmer received his final and most prestigious honor in early 1940, when he was named as one of two hundred people selected from over 3,000 nominees, to receive the Modern Pioneer of Science and Industry award. The banquet, held at the Palmer House Hotel in Chicago, honored fifty-nine midwesterners, but no one besides Elmer was from Rockford.[35]

As the honors piled up, his inventions and leadership ensured that the company would quickly leave the Depression behind and thrive. As the company grew in size, it began to acquire some unusually bright and committed engineers and other employees. Irl was completely in charge of the day-to-day running of the business and, thanks to the company's prosperity, he could begin to develop the beginnings of an innovative company outlook.

One of the foundations for that outlook was Irl's passion for consensus which, when combined with his enormous respect for the individual, led him to seek out like-minded people for employment. His management style in the late 1930s was to set the tone

Certificate from the American Society of Mechanical Engineers (ASME) naming Elmer Woodward as a society fellow, in 1938.

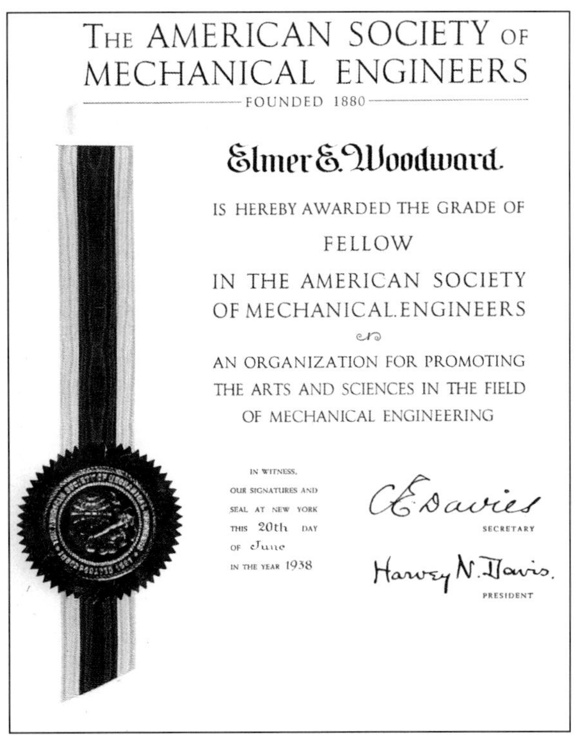
THE AMERICAN SOCIETY OF MECHANICAL ENGINEERS
FOUNDED 1880

Elmer E. Woodward

IS HEREBY AWARDED THE GRADE OF
FELLOW
IN THE AMERICAN SOCIETY OF MECHANICAL ENGINEERS

AN ORGANIZATION FOR PROMOTING THE ARTS AND SCIENCES IN THE FIELD OF MECHANICAL ENGINEERING

IN WITNESS, OUR SIGNATURES AND SEAL AT NEW YORK THIS 20th DAY OF June IN THE YEAR 1938

C. E. Davies
SECRETARY

Harvey N. Davis
PRESIDENT

E. E. Woodward To Be Honored

Governor Firm Head Is One of 200 in U. S. on Modern Pioneer List

Rockford will be represented by a delegation of business and industrial leaders at a banquet tonight at the Palmer house in Chicago honoring selected "modern pioneers," including Elmer E. Woodward, president of the Woodward Governor company.

Woodward, one of the nation's foremost inventors of governors for the regulation of hydro-electric equipment, engines, airplane propellors, is one of 200 men selected as "modern pioneers" by Dr. Karl T. Compton, president of the Massachusetts Institute of Technology, from an original list of 3,000 nominees.

Scientific Advancement

Woodward was en route to Chicago from Florida by air today to attend tonight's banquet, sponsored by the Illinois, Wisconsin, and National Manufacturers association. Illinois, Indiana, Iowa, and Wisconsin "modern pioneers" will be honored at the Chicago meeting.

Those selected are being honored for their contributions to scientific advancement and improved standards of living.

The Rockford delegation will be headed by R. S. Perry, vice president of the Ingersoll Milling Machine company.

A brief outline of Woodward's contributions to the art of prime mover control follows:

1877, started inventing, designing, developing and manufacturing governors exclusively.

1899, invented, designed, developed and start manufacture of the first mechanical compensating type of water wheel governor.

1911, improved and started manufacture of a line of hydraulic governors for water wheels which is today recognized as the best equipment of its kind.

1917, invented and started manufacture of the first equipment permitting remote control of hydroelectric units.

1921, developed and started manufacture of the first successful motor driven governor head for hydraulic governors.

Designed Magnet Generator

1930, supervised the design, development and manufacture of the permanent magnet generator, which today is the leading method of governor head drives.

1932, supervised the design and manufacture of the cabinet type actuator governing equipment.

1933, invented the first successful hydraulic governor for internal combustion engines.

1934, invented, designed, developed and started manufacture of the first governor successfully applied to the control of the pitch of airplane propellers.

1935, supervised the design and development of the first synchrophasing system, which both synchronized and phased the propellers of two or more engines or turbines to eliminate vibration in ships.

"Modern Pioneer"

—Register-Republic photo
E. E. WOODWARD

Elmer Woodward—a Modern Pioneer. This 1940 newspaper article, from the Rockford Register-Republic, *describes Elmer's selection by Massachusetts Institute of Technology (MIT) as one of 200 Modern Pioneers from around the country.*

and policies for years to come and would stress a "hands on" management philosophy in most areas. He was most comfortable with familiar things and basic values, where he could take the "high ground, morally and ethically."[36] Although he had some practical engineering skills, he was not much of an engineer. But he was a great organizer and salesman, and he was smart enough to leave the engineering to others like Willard J. (Bill) Whitehead, Luke Avery, Gil Gillespie, George E. (Bud) Sorensen, Urban Floor and, after Al Kalin left in 1938, G. Forrest (Forrie) Drake, who was put in charge of all engineering. Norm Noling, also one of this group, noted that these engineers were "the smartest all-around guys I've ever known."[37] Besides carrying on the sound engineering principles of Elmer, this group brought professional standards to the company and set the stage for the continued expansion of the company during and after World War II. No single one of them could ever hope to dominate engineering as Elmer had for generations, but as a group they were unsurpassed. Luck certainly played its part, because governor and control theory was complex, yet Woodward's domination of the field signaled a superior competence.[38]

Irl Martin was right in the middle of virtually everything that occurred at the company during this time. Its prosperity allowed him to try to make every aspect of the company professional. One of his major concerns was to insure that the best people possible were hired and placed in jobs appropriate to their abilities. He also encouraged those hired to learn many jobs, and to advance in the company. Moreover, he believed, the process of hiring and promotion should be more scientific. Accordingly, in 1938 Irl hired the

Psychological Corporation of New York City to do a psychological survey of the company's employees in order to make sure they were properly placed and to establish norms for the hiring of new personnel. Until this innovation encouraged other changes, the company had no professional personnel office or policies, no medical examination or medical insurance, and no vacation policy. The only benefit the workers got, besides being part of the new national social security system, was that each had a company-paid $500 life insurance policy.[39]

Engaging the Psychological Corporation was quite an unusual step, but Irl had heard a speech about their consulting abilities and was determined to try out their recommendations. Psychological consulting was a relatively new field in the 1930s, and the Psychological Corporation had developed an excellent reputation. Accordingly, Dr. Richard Schultz was sent out from New York to evaluate the skills, aptitudes, interests, and personalities of Woodward's personnel. After Schultz hired Leo Powelson as his clerk-stenographer, the battery of tests began. It included one- and two-handed coordination exercises on a simulated lathe, use of tools and instruments, and numerous paper and pencil tests on mechanical aptitude, general intelligence, spatial relations, and so forth. Schultz also interviewed each employee, and gave written and oral reports to Irl about the results. It took about the last nine months of 1938 to test everyone at the plant. All of this data formed the first detailed personnel records of the company.[40]

Dr. Schultz's recommendations cut both deep and wide, and one of them went against the grain of one cherished element in the company's ethics. Schultz had evaluated about a dozen of the 180 employees as having poor potential, and he recommended their termination from the company. Irl, who disliked firing people, broached the subject to Elmer. Elmer said that no one should be fired, and that was the decision that Irl followed. After Dr. Schultz's staff finished its assignment, testing continued under Leo Powelson's direction. The use of the Psychological Corporation to improve the company's professional performance turned out to be a competitive advantage, and spread Woodward's reputation as a progressive company throughout Rockford, attracting many young, confident people to the company.

Elmer Woodward and George Sorensen in 1939.

Although Woodward's starting wages were lower, they were able to hire people away from other companies in Rockford because their progressive image and cash profit sharing incentive were real inducements to loyal service, good workmanship, and esprit de corps.[41]

As Irl continued to transform the company and its property throughout the late 1930s, Woodward became a special place to work. The old line shafting came down and finer machines were purchased. Individual machine motors made the line shafting obsolete. The new machines had the closer tolerances necessary for precision diesel and aircraft work. Once the inside of the shop looked better, everything was painted and polished and kept as clean as possible. One continuing problem was that, although the shop was relatively neat and clean, it was not air conditioned. Irl held a plant meeting to discuss this issue and told the assembled group that he wanted air conditioning—not just for the fifth floor offices but for the whole plant. The rationale was partly for comfort but, more importantly, it was for cleanliness and better workmanship from drawings to machining. Irl asked the group for their wisdom and noted the project would cost about $100,000 and that to spend this much money would decrease bonuses that year. A vote was held, and the plant was air conditioned.

After air conditioning the plant, Irl promoted another innovative project. He had believed for a long time in the dignity of the worker and his work, and he especially believed that machinists had inferiority complexes when they compared themselves to other, more traditionally prestigious, occupations in medicine, law, or education. Irl emphasized how valuable their skills were, and he sought to instill pride in everyone. This led to his decision to require everyone to come to work clean-shaven and to wear clean shop coats while they worked. The rationale was simple: employees who looked good would feel good about themselves and would impress potential customers.[42]

These innovations were mostly complete by the end of 1939, and they were the first of many that would help transform the company in the 1940s. Sales were booming; plant capacity was getting overcrowded and, with the outbreak of World War II in September 1939, the company's products became vital to the nation's defense.

The Impact of World War II

1940-1945

On the last day of 1940, after sixty-three years of service and leadership and decades of valuable inventive work, Elmer Woodward died from a heart attack. He had been working all day at the office and passed away shortly after returning home. His death was an important milestone in the company's history, for the company was in the midst of a transformation brought about by new products, internal reorganization, and plans for expansion into a new building. Moreover, by this time, Irl Martin had been running the company for over a decade with Elmer's confidence and approval, and increasingly with independence. Irl also possessed an incomplete, if detailed, new view that the Woodward Governor Company ought to be organized and run differently than the typical American corporation. Some of these ideas would evolve from past company practices, and some would develop naturally from Elmer's example, but most would come about from a corporate vision that was slowly taking shape in Irl's mind.

This view was linked to Irl's abiding belief in capitalism, in its emphasis on individual initiative, and in the prosperity it could bring to people who worked hard. He also saw how individuals and companies had abused capitalism and how abuse had led to increased government intervention. For him, the New Deal was a prime example of this intervention, and Irl feared that, in the long term, people and business would grow more dependent on a bigger and more involved government. This trend, he believed, was unnecessary for people affiliated with the Woodward Governor Company, whatever good it may have done for the country at large. Shortly after the end of World War II, Irl was ready to formalize his business philosophy which had been taking shape over the last decade or so.

World War II would help transform the company, but Irl's evolving vision predated the war and would have come about without it. Nevertheless, the war did accelerate change within the company, just as it accelerated changes in American society. Some of these changes, such as the use of women in traditionally male jobs, were unforeseen and for a time resisted by society at large, Woodward included. Others, like the end of the Great Depression, were applauded by everyone who could remember the despair of the 1930s. Still, many of these same people, reflecting the disillusionment and isolationist feelings left over from World War I were wary of U.S. entry into another war, no matter how high the stakes. This latter group might support the America First Committee, as Irl Martin and a large number of Woodward employees did, and call for increased defense spending to make a "Fortress America." For a long while, they were adamantly opposed to aiding the Allies or to allowing the United States to get directly involved in Europe's or Asia's problems.

In Asia, China and Japan had been at war since 1937 and, in Europe, Hitler had acquired Austria and Czechoslovakia with hardly a shot fired. However, appeasement ended when Germany invaded Poland in September 1939, touching off the great conflagration of World War II. From then until the Japanese attack on Pearl Harbor on December 7, 1941, the United States was in an increasingly complex position between domestic demands for peace and security, and a growing perception that the Axis, if they won, might directly imperil the United States.

The struggle over what to do dominated the Roosevelt administration for the two years before Pearl Harbor. At first, the United States sold

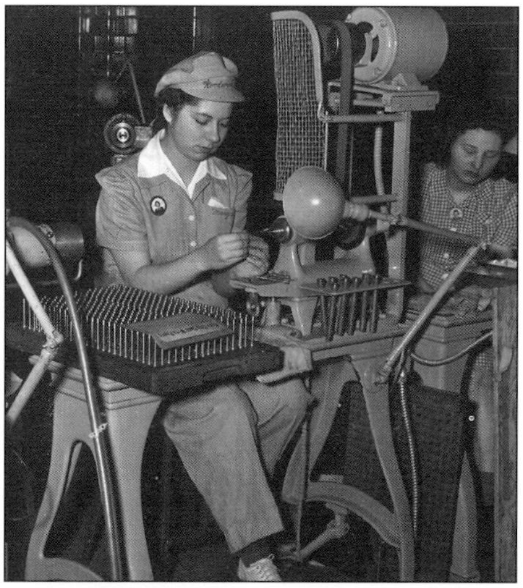

Because so many men left Woodward to serve in the Armed Forces, women were hired to fill in during their absence. These women are shown doing a variety of production jobs.

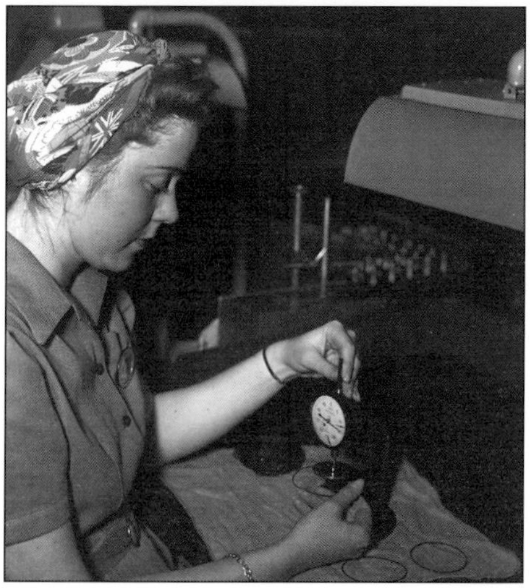

munitions to the Allies under a "cash-and-carry" plan. After Hitler had conquered France and most of Western Europe, thereby isolating Great Britain, the United States traded World War I era destroyers and ammunition for British bases in the Western Hemisphere. Then, in March 1941, the Lend-Lease Act virtually ended any pretense of neutrality, but it did allow the United States to stay out of the war while aiding Great Britain.

Just when it seemed Britain was in gravest danger, Hitler reversed course in June 1941 and attacked his ally, the Soviet Union. This blunder widened the war even more, but it took pressure off the British. Events by now were on a roller coaster. Japan had its eye on the oil- and rubber-rich South Pacific basin and Southeast Asia to add to its conquest of China. To complete its grand strategy, it would first have to eliminate the impediment of the U.S. Navy. The attack on Pearl Harbor was, at best, a partial success, however, since the Navy's aircraft carriers were at sea and escaped damage. Though a number of battleships and cruisers were lost, the attack actually backfired. It outraged America, galvanized the country's resolve to fight, and led to the organization of an allied force that included the United States, the Soviet Union, and Great Britain and her Commonwealth.

The war would go on for almost four more years with a ferocity and a death rate for soldiers and civilians unprecedented in world history. Slowly the tide would turn in the Pacific with the U.S. Navy's victory at the Battle of Midway, fought just six months after Pearl Harbor. In Russia, the German advance was halted at Stalingrad. In North Africa, at El Alamein, Morocco and Algeria, Allied victories began to chip away at Germany's empire. The war's outcome would

Rockford College students enrolled in a wartime work-study program punch in their Woodward time cards, sometime in the fall of 1942. The sign reminds them that "every piece of scrap hurts us and aids the Axis."

remain uncertain throughout 1943, despite Allied gains and a Soviet offense in the East. But, in 1944, with the D-Day landings in Normandy, the end was in sight. In the meantime, Japan was in slow retreat everywhere from the islands of the South Pacific to the mainland of Asia. By 1945, both Axis nations were on the verge of defeat, which finally came with Germany's unconditional surrender in May and, after two atomic bombs, Japan's surrender in August.

During this time, industrial capacity in the United States grew rapidly. Full employment was reached in 1943—an incredible turnaround from the Depression lows of the previous decade. And, of all the major belligerents of World War II, only the United States emerged physically unscathed.

Rockford, like the rest of the nation, made sacrifices for the war effort. Rationing, blackouts, scrap drives, victory gardens, and increased working hours (and higher pay) became the norm. Patriotism was appealed to, though with fewer First Amendment restrictions. Gold Star mothers, whose sons had died in battle, became the nation's heroines and the objects of great sympathy. Tales of individual and group heroism became part of the public folklore. Fascism was reviled, and our Allies, even the Soviet Union, were praised. It was an all-out effort by the people of the United States to win the war as quickly as possible.

The Woodward Governor Company was a microcosm of this effort. Despite its small size, it made important contributions to winning the war primarily because its governors were applied on new ships and planes. Sales of the diesel and airplane propeller governors boomed throughout the war and brought a new level of prosperity to the company, even after heavy taxes, renegotiated contracts, and voluntary reductions of prices.

In the years before the war, the company moved toward recapitalization to make its stock more available to the public. The majority of the 300 shares of common stock were owned by Elmer Woodward and his extended family. Elmer, while vigorous, was in his seventies, and had turned down the offer of a group of East Coast people to buy the company at *any* price Elmer thought the company was worth. To Elmer, the offer was troubling. He had told the buyer's representative that making money was not his personal objective, and that he felt a responsibility to the people who worked for Woodward. Under new ownership, he was worried that the workers might not be protected in the way he wanted them to be. The offer was tempting, though. Elmer could have guaranteed security for himself and his family for a long time to come, but he chose to continue to control the company in a way he believed was beneficial to all the company's employees.[1] This was heartening, and unusual, since many family companies had been sold or liquidated in the 1930s. But it was also troubling because of the uncertainty over how long Elmer could retain control due to his advancing age. Also, while the company was in sound financial condition, new tax laws dictated a change in capitalization in order to protect the original stockholders' interests. Therefore, the decision was made on November 1, 1939, to proceed with a recapitalization plan that would, in less than a year, encourage Woodward employees and the general public to become shareholders of the company.[2]

By the summer of 1940, the new capitalization plan allowed Woodward employees first and then outsiders to become stockholders in the company. Woodward stock became a free market stock for the first time although, compared to most companies, it was still rather closely held. Irl had lost money in the great stock market crash that began in 1929, and he was wary of anything that appeared to be speculation. However, Elmer had for a long time wanted to reward employees by allowing more of them to become part of the "other side" of the company. The plans to allow employee purchases were conservative and so were the details of the recapitalization plan.

PROSPECTUS

Woodward Governor Company

62,000

Common Shares

$1 Par Value

THESE SECURITIES HAVE NOT BEEN APPROVED OR DISAPPROVED BY THE SECURITIES AND EXCHANGE COMMISSION

Woodward Governor Company has registered the securities by filing certain information with the Commission. The Commission has not passed on the merits of any securities registered with it.

IT IS A CRIMINAL OFFENSE TO REPRESENT THAT THE COMMISSION HAS APPROVED THESE SECURITIES OR HAS MADE ANY FINDING THAT THE STATEMENTS IN THE PROSPECTUS OR IN THE REGISTRATION STATEMENT ARE CORRECT.

	Price to Public	Underwriting Discounts or Commissions	Proceeds to Company*	Proceeds to Certain Shareholders*
Total	$837,000	$124,000	$115,000	$598,000
Per Unit	$13.50	$2	$11.50	$11.50

*Of the 62,000 Common Shares offered hereby, the offering of only 10,000 shares represents financing on the part of the Company. The remaining 52,000 shares are outstanding and owned by certain shareholders of the Company and the Company will not receive any proceeds from the sale of such 52,000 shares. The names of such shareholders and the respective numbers of shares agreed to be sold by them to the several Underwriters are set forth herein under the heading "Underwriting Agreement".

These Common Shares are offered subject to prior sale and to withdrawal, cancellation or modification of the offer without notice, to delivery to and acceptance by the Underwriters named below, and subject to the approval of the legality of issuance and certain other matters by Messrs. Pam, Hurd & Reichmann, counsel for the Underwriters named below, and by Messrs. Bobb Spoerri Bourland & Harris, and Charles S. Thomas, Esq., counsel for the Company. The right is reserved to reject any and all orders.

TO FACILITATE THE OFFERING, IT IS INTENDED TO STABILIZE THE PRICE OF THE COMMON SHARES OF THE COMPANY. THIS STATEMENT IS NOT AN ASSURANCE THAT THE PRICE OF THE ABOVE SECURITY WILL BE STABILIZED OR THAT THE STABILIZING, IF COMMENCED, WILL NOT BE DISCONTINUED AT ANY TIME. IN THIS CONNECTION, SEE "TERMS OF OFFERING" HEREIN.

FOR A STATEMENT AS TO AGREEMENTS BETWEEN THE COMPANY, CERTAIN OF ITS SHAREHOLDERS AND THE UNDERWRITERS, RESTRICTING THE RIGHTS OF THE COMPANY AND SUCH SHAREHOLDERS TO PURCHASE, SELL OR OTHERWISE DISPOSE OF COMMON SHARES OF THE COMPANY, SEE "UNDERWRITING AGREEMENT" HEREIN.

PAUL H. DAVIS & CO. **RAWSON LIZARS & CO.** **OTIS & CO.**
AMES, EMERICH & CO., INC. **THE FIRST CLEVELAND CORPORATION**
DEMPSEY-DETMER & CO. **FULLER, CRUTTENDEN & CO.**
CARTER H. HARRISON & CO. **ROGERS & TRACY, INC.**

No dealer, salesman or other person has been authorized to give any information or to make any representations, other than those contained in this Prospectus, in connection with the offer contained in this Prospectus, and if given or made, such information or representations must not be relied upon as having been authorized by the Company or by any of the Underwriters.

The date of this Prospectus is July 30, 1940

The prospectus for the first public offering of Woodward Governor Company stock, dated July 30, 1940.

The Board of Directors amended the articles of incorporation of the company. This allowed them to change the 300 common shares having $100 par value to 240,000 common shares having $1 par value. Each shareholder received 100 of the new $1 par value shares for each share held before November 1, 1939. To make this more equitable, a dividend of $5 a share (paid in stock) was distributed to the shareholders and $150,000 was transferred from the earned surplus of the corporation to stated capital. Thus, the new corporation had a stated capital of $180,000 and 180,000 common shares were issued. Another 60,000 common shares were authorized and held in reserve should the company want to make them available for purchase.[3] The impact on the shareholders was considerable. Estelle Milne and Thomas Davis, the smallest shareholders, each saw their 2 shares jump to 1,200. Irl Martin's 20 shares became 12,000 and Elmer Woodward's 98 shares became 58,800.[4]

Other members of the Woodward family, including Irl's wife Dorothy, held thousands of shares, but the largest non-family block of 25,800 was owned by Robert and Alice Kelley. Robert Kelley had been with the company since the turn of the century and had served in a number of executive positions, including vice president. Poor health in the 1930s, however, had led to his semi-retirement from company affairs and his decision to spend winters in Florida. He and Elmer remained friendly, and he often served as a host for the golfing vacation trips that Elmer and George Sorensen made to avoid the midwestern winters. Irl believed that these trips were necessary to preserve Elmer's health and vigor.

Before their trip scheduled for January and February 1940, Elmer and George had worked out a deal with Kelley to buy all of his shares in the company, at an agreed-upon price of $10 per share. The idea was then for Elmer to resell the shares to company members for the same price. When Elmer and George got to Florida, however, they were surprised to learn that Kelley had

Irl Martin, his wife Dorothy, and Elmer Woodward vacationing in Florida in 1938.

jumped the price to $10.83 per share. Elmer was taken aback by this sudden change. In a letter to Irl, he complained that Kelley, at the very least, owed some "consideration" to "the boys" in the shop whose hard work had made his retirement years wealthy and comfortable. Besides, Elmer added, his wife had inherited a tidy sum herself, and Bob didn't need the extra money.[5]

The letter expresses a profound sense of exasperation, and some anger. In one passage, Elmer bitterly announced that Kelley would lose his respect and all ties with the company unless he returned to the original price.[6] But the bitterness and anger soon subsided. Kelley remained a friend.

Elmer and George continued on with their golfing vacations, Elmer staying as a guest in the Kelley home, George residing at a nearby hotel so as not to impose on the Kelleys.[7] And, to top it off, Elmer not only agreed to the $10.83 price, but he still sold the stock to employees at $10 a share, making up the difference ($21,414) himself.

Thanks to his bad experiences during the 1929 Crash, Irl had learned to be wary of stock purchases in general. Elmer, on the other hand, wanted company employees to purchase stock at a price *lower* than the market value. The purpose was not to offer extra compensation, but to give them a stake in the company. Thus, the announcement was made offering them stock at $10 a share. The announcement also noted that a bank official would counsel anyone who wanted advice on how many shares he could reasonably afford. The official was Eugene Abegg, perhaps Irl's best friend, and his advice was paternalistic and conservative. Employees were quizzed about their financial status and then told how many shares they should buy. Leo Powelson, for example, who began work in 1938 for $50 a month but who had received salary increases and profit sharing since then, was told that he could buy forty shares and was loaned $400 by Abegg's bank to make the purchase.[8]

The offer was immensely popular. On April 17, 1940, 81 of the 306 employees bought a total of 6,350 shares.[9] The average individual purchase was less than 80 shares, but the price was a bargain. With this transaction, Elmer's wish to share his closely-held company with its employees was granted.

Once this transaction was completed, the company moved to offer its stock to the general public through an agreement with nine underwriters led by Paul H. Davis & Company of Chicago. The Davis Company hired the Day & Zimmermann Engineering Company of Philadelphia to make a detailed report on the company for the purpose of evaluating its current worth and future prospects. The report's favorable conclusion noted the company's leadership in products and its worldwide reputation. It also highlighted the company's aptitude testing program, its good facilities, and the "highly satisfactory" employer-employee relationship which seemed to preclude labor difficulties.[10] The underwriters were impressed and developed a prospectus to sell 62,000 shares of the company to the public.

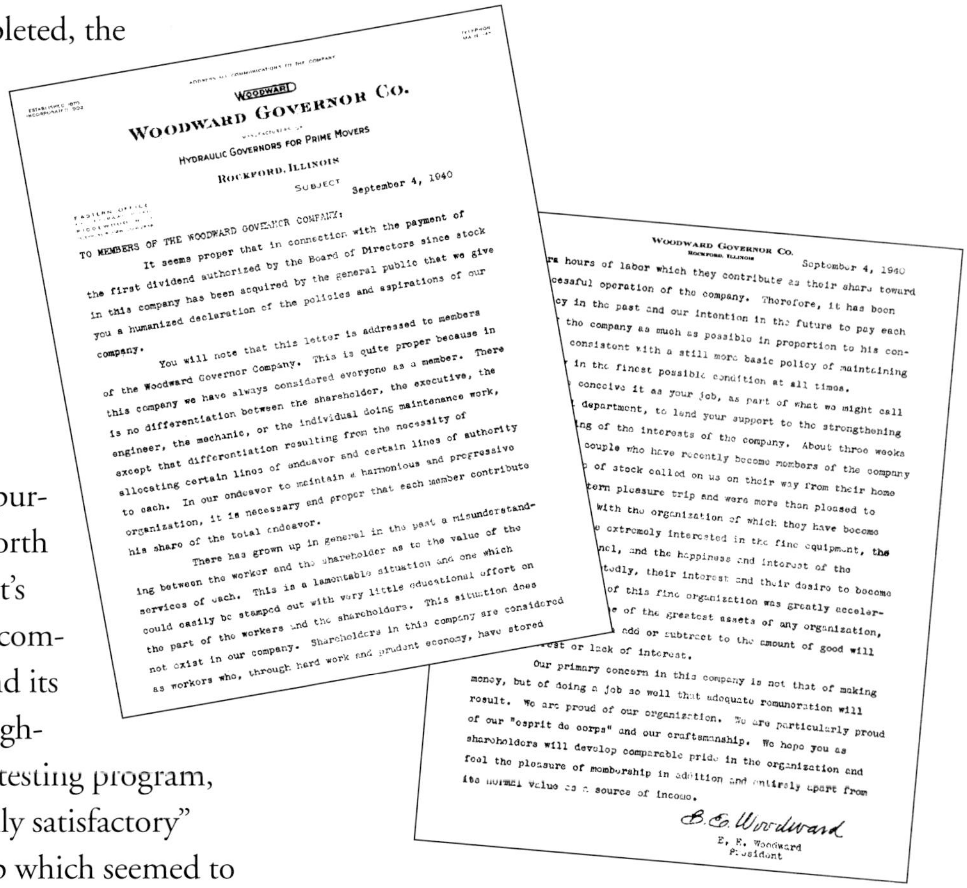

WOODWARD GOVERNOR CO.

HYDRAULIC GOVERNORS FOR PRIME MOVERS

ROCKFORD, ILLINOIS

SUBJECT September 4, 1940

TO MEMBERS OF THE WOODWARD GOVERNOR COMPANY:

It seems proper that in connection with the payment of the first dividend authorized by the Board of Directors since stock in this company has been acquired by the general public that we give you a humanized declaration of the policies and aspirations of our company.

You will note that this letter is addressed to members of the Woodward Governor Company. This is quite proper because in this company we have always considered everyone as a member. There is no differentiation between the shareholder, the executive, the engineer, the mechanic, or the individual doing maintenance work, except that differentiation resulting from the necessity of allocating certain lines of endeavor and certain lines of authority to each. In our endeavor to maintain a harmonious and progressive organization, it is necessary and proper that each member contribute his share of the total endeavor.

There has grown up in general in the past a misunderstanding between the worker and the shareholder as to the value of the services of each. This is a lamentable situation and one which could easily be stamped out with very little educational effort on the part of the workers and the shareholders. This situation does not exist in our company. Shareholders in this company are considered as workers who, through hard work and prudent economy, have stored

WOODWARD GOVERNOR CO.

ROCKFORD, ILLINOIS

September 4, 1940

...rs hours of labor which they contribute as their share toward
...cessful operation of the company. Therefore, it has been
...cy in the past and our intention in the future to pay each
... the company as much as possible in proportion to his con-
... consistent with a still more basic policy of maintaining
... in the finest possible condition at all times.
... conceive it as your job, as part of what we might call
... department, to lend your support to the strengthening
...ng of the interests of the company. About three weeks
... couple who have recently become members of the company
...p of stock called on us on their way from their home
...tern pleasure trip and were more than pleased to
... with the organization of which they have become
...e extremely interested in the fine equipment, the
...nel, and the happiness and interest of the
...tedly, their interest and their desire to become
... of this fine organization was greatly acceler-
...e of the greatest assets of any organization,
... add or subtract to the amount of good will
...rest or lack of interest.

Our primary concern in this company is not that of making money, but of doing a job so well that adequate remuneration will result. We are proud of our organization. We are particularly proud of our "esprit de corps" and our craftsmanship. We hope you as shareholders will develop comparable pride in the organization and feel the pleasure of membership in addition and entirely apart from its normal value as a source of income.

E. E. Woodward

E. E. Woodward
President

This letter, signed by Elmer Woodward in 1940, contains one of the first expressions of Irl Martin's Corporate Partnership philosophy.

Of these, 52,000 were from company stockholders, most notably Elmer Woodward, who sold 22,800 of his shares. Other Woodward family members sold a part or all of their shares, but Irl Martin kept all of his 12,000 shares. The other 10,000 came from the 60,000 other shares the company held in reserve. After all was said and done, the company would have 190,000 shares in circulation and another 50,000 unissued. Initially Elmer and Irl believed the stock was worth $15 a share, but the underwriters agreed to offer it for $13.50 a share. After the underwriters' fees of $2 a share, the net proceeds per share were $11.50, for a total gain of $115,000.[11] This sum went into the company funds, while the rest of the net proceeds ($598,000) remained the personal funds of the shareholders who sold them. These transactions taken altogether thus increased the number of stockholders from seventeen into the hundreds, although the Woodward family still held almost 60 percent of the issued stock.

By August 1940, the Securities and Exchange Commission approved the stock sales plan and all of the 62,000 shares were sold through the underwriting syndicate. On August 12, the company's Board of Directors met and declared a dividend of twenty-five cents per share on the 190,000 shares then in circulation. This was the fifty-third dividend since incorporation in 1902 and the first under the recapitalization.[12]

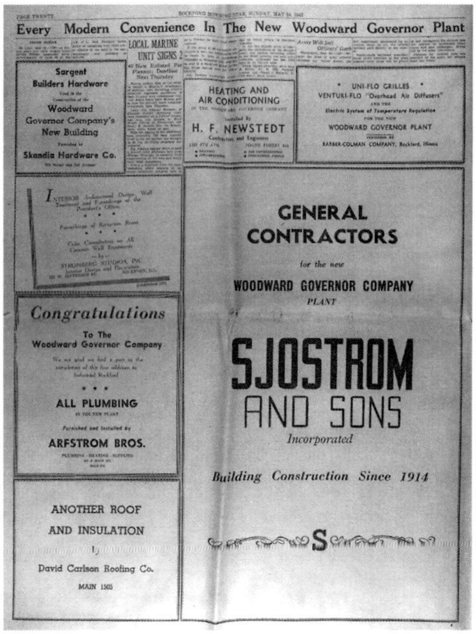

Every Modern Convenience In The New Woodward Governor Plant

LOCAL MARINE UNIT SIGNS

Sargent Builders Hardware Used in the Construction of the Woodward Governor Company's New Building Furnished by Skandia Hardware Co.

HEATING AND AIR CONDITIONING H. F. NEWSTEDT

UNI-FLO GRILLES VENTURI-FLO "Overhead Air Diffusers" WOODWARD GOVERNOR PLANT BARBER-COLMAN COMPANY, Rockford, Illinois

GENERAL CONTRACTORS for the new WOODWARD GOVERNOR COMPANY PLANT

SJOSTROM AND SONS Incorporated

Building Construction Since 1914

Congratulations To The Woodward Governor Company

ALL PLUMBING Furnished and Installed by ARFSTROM BROS.

ANOTHER ROOF AND INSULATION David Carlson Roofing Co. MAIN 1505

Products Of Woodward Governor Company Are World Famous

INVENTOR LAID FOUNDATION OF WOODWARD CO.

MAY RESCIND CUT IN INCOME TAX EXEMPTION

MARQUETTE CEMENT Used in the Construction of The New WOODWARD GOVERNOR COMPANY Plant Furnished by J.H. PATTERSON CO.

WOODWARD GOVERNOR Another Rinehimer Installation of Fine Architectural Woodwork RINEHIMER BROS. MFG. CO.

EXCAVATING AND GRADING ANDERSON BROS.

Plain and Ornamental PLASTERING In the New Woodward Plant by GRIP BROS.

Congratulations To Woodward Governor Company Structural Steel Furnished and Erected By A.C. WOODS & CO.

4,000 LBS. of COPPER MASTER SHEET METAL AND FURNACE SHOP

BURDICK ULTRAVIOLET SOLARIUM THE BURDICK CORPORATION

Car-Pet-Line Congratulates WOODWARD GOVERNOR COMPANY ACOUSTICAL CEILINGS and FLOOR COVERINGS The CAR-PET-LINE STORE, Inc.

Newspaper article and congratulatory advertisements, dated May 24, 1942, featuring Woodward's new North Second Street plant.

Enclosed with the dividend checks, mailed out in early September, was a letter addressed "To Members of the Woodward Governor Company." The letter, signed by Elmer, was the idea and handiwork of Irl Martin. It expressed an opinion long held by Elmer and Irl and put into practice before, but never explicitly stated on paper. For the benefit of the general public and for posterity, Irl emphasized the company's unique view of the relationships among wage earners, salaried personnel, executives, and stockholders. He sought to differentiate Woodward from other companies by emphasizing mutual interests and teamwork among *members,* not employees or shareholders. The most significant parts of the letter express a humanistic business philosophy years ahead of its time:

It seems proper that in connection with the payment of the first dividend authorized by the Board of Directors since stock in this company has been acquired by the general public that we give you a humanized declaration of the policies and aspirations of our company.

You will note that this letter is addressed to members of the Woodward Governor Company. This is quite proper because in this company we have always considered everyone as a member. There is no differentiation between the shareholder, the executive, the engineer, the mechanic, or the individual doing maintenance work, except that differentiation resulting from the necessity of allocating certain lines of endeavor and certain lines of authority to each. In our endeavor to maintain a harmonious and progressive organization, it is necessary and proper that each member contribute his share of the total endeavor.

...Shareholders in this company are considered as workers who, through hard work and prudent economy, have stored up extra hours of labor which they contribute as their share toward the successful operation of the company. Therefore, it has been our policy in the past and our intention in the future to pay each member of the company as much as possible in proportion to his contribution, consistent with a still more basic policy of maintaining the company in the finest possible condition at all times.

Irl concluded the letter by calling on shareholders to aid the "esprit de corps" of the company by getting involved and promoting good will—"one of the greatest assets of any organization." He also pointed to the traditional company goal of making the best possible products, because by doing so "adequate remuneration will result."[13] This concept of membership and service was a major building block of the Corporate Partnership and Woodward Way ideas that Irl developed and elaborated upon during and after World War II. By the end of the war, Irl would put these ideas and others into a plan for corporate life that still guides the company.

As the war raged in Europe during the summer and autumn of 1940, demand for the company's adjustable pitch airplane propeller governors seemed inexhaustible. Hamilton Standard propellers with Woodward governors were found, among other places, on the Spitfires and Hurricanes that won the Battle of Britain and staved off Hitler's planned invasion.

Because of this great demand for propeller governors, and strong demand as well for diesel and water wheel governors, the company went from 300 to 400 members between March and October 1940, and was running out of manufacturing space at its Mill Street complex. New machine tools and plant equipment were on order and more were needed. The circumstances dictated a move to a new location since there was not enough land to expand the old plant. Accordingly, in early November, the company purchased for $30,000 a twenty-five acre site on North Second Street adjacent to Loves Park, a Rockford suburb. This was the site of the old interurban car barn, a concrete building with 20,000 square feet of floor space, and plenty of additional land to erect a new plant. Eventually all operations would be moved there. In the meantime, however, the car barn was refurbished and some machines were moved to the new location.[14]

By the time the annual meeting was held on December 5, 1940, the company was on the verge of completing its most successful and potentially revolutionary year. Plans for the new company facility north of Rockford were elaborated, and the decision to spend $800,000 for the plant and additional equipment was ratified. The company had no alternative since unfilled orders had risen to almost $4.4 million as compared to $1.7 million just eight months before.[15]

These were euphoric, if tense, times. Planning a new building, beginning operations at the new site, and keeping quality and production up at both locations posed problems no one in the company had faced before. In the midst of this, on December 31, Elmer Woodward died of a heart attack. Elmer was eulogized in death as he had been honored in life. Reverend John Gordon, later to be the company chaplain, noted, "He fulfills the definition of a gentleman as one who puts more into life than he takes out. He was interested in every good thing in our city and his loss will be mourned by multitudes of people." A fellow Rotarian described him by saying, "I never knew a more considerate or modest man," and the Rotary Club passed a resolution that noted his inventive genius and international reputation and concluded:

But today we are thinking of the charm we felt in his presence, the charm we now so greatly miss. Elmer Woodward will live forever in our memories as an ideal to be emulated—an ideal beyond the capacity of most of us, but he is an inspiration to thorough mastery of our vocation, and to kindly living.[16]

The company would also miss Elmer's tempering ability. He often had refused to make a decision until he knew all the facts—a trait that initially frustrated Irl Martin's impatient spirit. Irl admired this trait and sought to emulate it but, in the 1930s, he had sometimes gone home angry and frustrated because "the old man" would walk out of the room rather than make a decision that Irl wanted. However, when Elmer came back the next day, he knew the facts and the decision would come easily.[17]

Elmer's death had a profound impact on Irl Martin, who had expressed the kind of love a son could genuinely feel for a father. A year after Elmer's death, Irl described him as "the finest man I have ever known," one who had "all the qualifications you could have." He was "very human" and abhorred the strife too common between management and labor. He worked harder than anyone in the company because "work was his pleasure and hobby." Money was the necessary by-product of his success but not his major purpose in working. Hence, when his people got in trouble, "without fanfare" he helped them as one would do to any member of his family.[18]

With Elmer's death, the burden and the glory of running the company without fear of being overruled was Irl's. Irl was up to the challenge and, through good health and bad, he would dominate the company for thirty-five more years as president until 1960, and as chairman of the board until his death in 1976.[19] During all this time, his model remained Elmer.

Elmer left a handsome estate valued at almost three quarters of a million dollars—the vast majority of which had been accumulated in the 1935-1940 period. The bulk of the estate was his 38,000 shares of company stock worth $460,000 and bank balances of almost $150,000. He also owned savings bonds, street improvement bonds, common stock and promissory notes for money loaned to family members. Because Elmer's wife had died over a decade before, his entire estate was divided equally among his three children, Dorothy C. Martin, Harold, and Walter. This made Dorothy a larger stockholder than her husband, Irl, and it also brought Elmer's two sons back into the company picture, although neither would play a very active part in the company's daily activities. Walter continued serving the company as a vice president and member of the Board of Directors. Later, after Walter left the board, Harold (Doc) took his place. Eventually both left the company for other interests.[20]

Part of the large influx of new members during World War II, this 1943 "induction class" receives an orientation tour from Harold "Doc" Woodward (far left), Elmer's son.

By the time Elmer died, Irl had led the way in promoting a business philosophy that was elaborated upon in the company's first *Introduction Manual for Applicants,* published on December 16, 1940. This manual underlined the use of the term "member" rather than employee or worker when discussing people affiliated with the company. It also noted benefits members received, such as paid vacations given at the rate of one day for each year of service up to a maximum of two weeks, a free yearly medical exam, life insurance up to a maximum benefit of $1,000, and a work schedule between forty and fifty hours aimed at preventing layoffs during cycles of receding business. Interestingly, cash bonus payments were not mentioned despite their having been paid on a yearly basis, but they were hinted at in the statement: "It is the company's policy in periods of good income to pay out in wage-adjustments that part of its income above reserve requirements."[21]

Because the company defines itself as a family, the manual included some benevolently parental prohibitions against wasting money, loans between individual members, intemperance, wage assignments and garnishments, family difficulties, and immoral behavior. It noted that, while it was not the company's practice to loan money to members, in an emergency it "will consider it a privilege to supply any of its members with temporary help." It also noted that failure to follow the other generally accepted societal practices noted above would "subject members to immediate dismissal." A dress code was also established, requiring machine operators to wear bow ties and shop coats with half-length sleeves and requesting all members to come to work clean-shaven and neatly dressed in order to create "self-respect and efficiency."[22]

To carry out these and other policy requirements, as well as to have a mechanism for changing policies, a "multiple management" system similar to that developed by the McCormick Spice Company of Baltimore was adopted by Woodward. This system was composed of three committees, a Senior Committee, a Junior Committee, and a Primary Committee. In theory, they were modeled after the three branches of the American government, but the system as a whole was more restrictive, and an executive veto could not be overturned. The Primary Committee was like the House of Representatives and was made up of twenty-five non-supervisory members. It made recommendations to the Junior Committee so long as twenty of the twenty-five members agreed. The Junior Committee was modeled after the Senate and consisted of fifteen junior executives or supervisors. It had to vote unanimously for a recommendation from the Primary Committee to go to the Senior Committee. The Senior Committee, the executive branch, consisted of the company officers and senior executives. They would then decide whether to accept the recommendation.[23] In reality, it meant that Irl Martin had to agree to it, since his veto could not be overturned. On the other hand, he was educable and interested in company harmony. Therefore, he did not often seek to block recommendations that had near unanimous support. This system lasted until 1945 when, because of perceived inefficiencies, it was replaced by the Vertical Committees that the company has used ever since. Still, the multiple management system was innovative, and it did alleviate some of the pressure that Irl, as president, felt during the company's expansion in 1941.

Irl Martin meeting with the Senior Committee in his office in 1942. Shown (left to right) are: Roy Estell, Walt Thorell, Forrest Drake, and Charles Turner.

The United States moved from neutrality to nonbelligerent support of the Allies during 1941. By autumn the country was waging economic war against Japan, and American sailors and merchant men were dying in an undeclared naval war in the North Atlantic. These actions led to a renewal of the Selective Service Act and a rapidly expanding national defense

The Primary Committee annual dinner for 1944, held at the Wagon Wheel in Rockton. The Primary Committee was a forerunner of the Vertical Committees established after the war.

TABLE THIRTEEN

Sales Year Ended September 30, 1941	
Water Wheel Governors	$ 745,210.60
Internal Combustion Engine Governors	845,369.27
Airplane Propeller Governors	2,968,060.64
TOTAL SALES	$4,558,640.61
Unfilled Orders (Approximately)	$7,800,000.00

program. As a result, Woodward's products were increasingly in demand even before the attack on Pearl Harbor, as the statistics at the end of the company's 1941 fiscal year show (see Table Thirteen).[24] Sales were more than double fiscal 1940 and unfilled orders had increased fourfold.

Besides the day-to-day task of keeping the operation running efficiently and making sure production continued to increase, Irl was preoccupied during 1941 with the decision to build a new facility. He did not want just a new plant; he wanted the finest industrial building in the world. He knew definitely what he wanted and, to fulfill his goals, he went to a trusted source, the Harza Engineering Company, to build the new plant. Harza had built power plants equipped with Woodward governors all over the world. Moreover, Irl was good friends with Eric Floor, an execu-

North Second Street plant under construction about 1941, shortly before completion. The south wing (to the right of the main entrance) was the former interurban car barn and was incorporated into the main construction. The separate building at the far right formerly housed a pickle works.

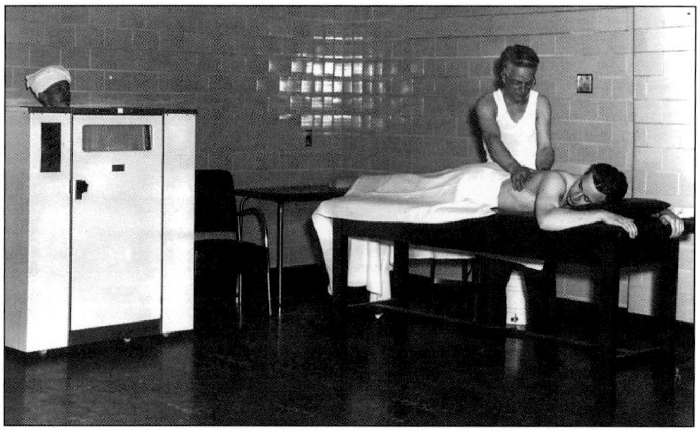

A 1942 description of the Second Street treatment room, reads: "Equipped with treatment table and infra-red heat cabinet, the facilities of this room are available to all members as required to alleviate minor aches and pains before they assume proportions detrimental both to member and company in time lost and impaired production." Shown are Leo Powelson in the cabinet, "Doc" Varney standing, and Harold Hopkins on the table.

tive of Harza, whose son Urban was an engineer at Woodward. Since Irl wanted a building that looked like a power plant, Harza was a natural choice.[25]

Construction continued throughout 1941, and the plant was finally ready for complete occupation early in 1942. The result was state-of-the-art for its time. It was fully air-conditioned, temperature and humidity controlled, and equipped with electric precipitators to filter out smoke, dust, and even bacteria. The machine shop had a terrazzo floor with walls of ceramic tile. The ceilings were finished with acoustic tile and fluorescent lighting. The basement contained a first aid room, a cafeteria, and locker rooms with showers. It also had Irl's favorite, a sun room designed so that a three-minute stay twice a week in the ultra-violet and infra-red rays would be the equivalent of healthful summer sunshine. There was also a 500-seat auditorium for company meetings and other functions.[26]

One morning while the plant was still under construction, the United Auto Workers union distributed a flyer at the Loves Park location. This was the first visible unionizing effort made at Woodward within memory, and it went nowhere.[27] It was Irl's belief that in a well-managed company like Woodward, unions were unnecessary, and no one in the company seemed to disagree.

Like many defense industries, the company did get financial advantages from the government due to the war emergency. Woodward spent $365,779.62 for the purchase of machinery required to manufacture products essential to the war effort. The federal government promised to reimburse the company through monthly payments spread over five years.[28] (Thus, while the new plant was constructed entirely with company money, some of the machinery was purchased through this federally sponsored plan.)

Another economic action in 1941 highlighted the emergence of some different men as key members of the company. This was the company's decision on April 15, 1941, to purchase and be the beneficiary for life insurance policies for Irl Martin, Forrest (Forrie) Drake, Ward Ring, Walter Thorell, Roy Estell, and Charles Turner. Drake and Turner were relatively new members who headed engineering and finance, respectively, and were rapidly rising in the corporate hierarchy. When Irl was officially named president on August 6, 1941, Charlie Turner was named assistant secretary and assistant treasurer.[29]

In September, the directors also agreed to provide annual bonuses. The amounts were based upon the period of service and company officers' estimates of each individual's value to the company.[30]

Later in the war, the bonus plan would be changed to be more systematic and more equitable.

After spending most of 1941 as a tandem director with Walter Woodward, Irl proposed a change in the company's bylaws to raise the number of directors from three to five. Although Irl was ill when the annual meeting was held on December 4, 1941, the shareholders or their proxies nominated and elected three new directors: Walter Thorell, the company secretary; George W. (Bill) Taylor, a member and Elmer Woodward's nephew; and Paul H. Davis, a friend of Irl's and the security underwriter who had helped the company go public in 1940.[31] The company was beginning to take on a new shape molded by Irl, even though he was not at the meeting.

Irl had entered a Rockford hospital on December 1, 1941, for observation and rest after being taken ill at the plant.[32] He remained out of company affairs for over two months before recovering. Although some members thought he had suffered a heart attack, he had, in fact, succumbed to a nervous breakdown. His recuperation was relatively slow but, when he did return, he was in vigorous health and as feisty as ever. In his absence, the United States had entered World War II, and the company its new plant.

Irl appeared before the assembled members in a March 1942 plant meeting. After thanking them for their best wishes during his illness he said, "I'll give you all the flowers to start with. Then I'll give you hell as usual." He then launched into an exhortation on the necessity of keeping the plant clean—one of his pet topics. Irl was a true believer in everyone looking good while working in the clean, attractive plant that was their new industrial home.

Jean Ebling in the lobby of the new plant, 1942.

Another topic covered at this same meeting was working women. Decades before the women's liberation movement, Irl echoed the prevailing belief that women were the bulwark and backbone of the family, and that their responsibility to society was to maintain the home and help raise the children. "I don't like the idea of women in a factory to run a machine. Their main place in this world is at home whether they like it or not. I believe that." But, he then went on, there was the war, and American men were leaving the factories to fight it.

The cafeteria line, 1942.

In 1942, fear that the war might come to American soil prompted many people to take precautions, and Woodward was no exception. These members of the plant protective force are taking target practice at a pistol range located behind the plant.

and he appeared to revel in the communication opportunity that the microphone gave him. He was clearly the boss, but he wanted to run the company by consensus. He was open to ideas, and he actively sought them, sometimes changing his mind instantly when convinced by a better argument. He spoke "horse sense," a favorite phrase, and often told his audience, "You are going to hear it whether you like it or not." He felt obligated to be the "father" of the company and the head cheerleader for its members, whom he called "the finest people" in the world. But after a buildup like this, he could also deflate his listeners by noting that he would "can half of the force" if the plant was not kept clean. This might be followed by a controversial idea that Irl promoted by saying, "I think that's only fair." When Irl thought it was "fair," everyone knew that he advocated it.

Sometimes Irl was repetitious and condescendingly tolerant, and sometimes funny. His statements were straightforward, and he was unconcerned about their consequences. Still, he always had an overriding concern for the welfare of the company and of the individuals who made up its

Like it or not, women had to take their place, and Woodward was no exception. Irl was not complaining; he was simply noting the truth. Thus, the company hired women, and Irl promised that they would be treated as fairly as any regular member. Then, Irl concluded, "When the war is over, they can go back home."[33] It was a blunt talk by a blunt man with no hesitation to express his opinions. But bluntness did not dull the edge of his fairness or humanity. In fact, this traditional reaction only reflected the attitude of the federal government which later commissioned films showing women eagerly promising to give up their jobs to returning service men so they could get married and begin raising families.

Irl's plant meetings became a tradition during World War II,

The Machine Shop and mezzanine in 1942. The banner partially visible behind the American flag reads "KEEP 'EM FLYING!"

membership. He was a good psychologist when he promoted togetherness, loyalty, integrity, and patriotism. Most of all, he was a confident leader who kept his word.[34]

While Irl consistently believed that income taxes and excess profit taxes were too high and stifled initiative, he patriotically supported the war and made sure that Woodward's prices were reasonable. As one example, on May 15, 1942, the company unilaterally reduced prices 20 percent on its airplane governors when it sold them to Hamilton Standard.[35]

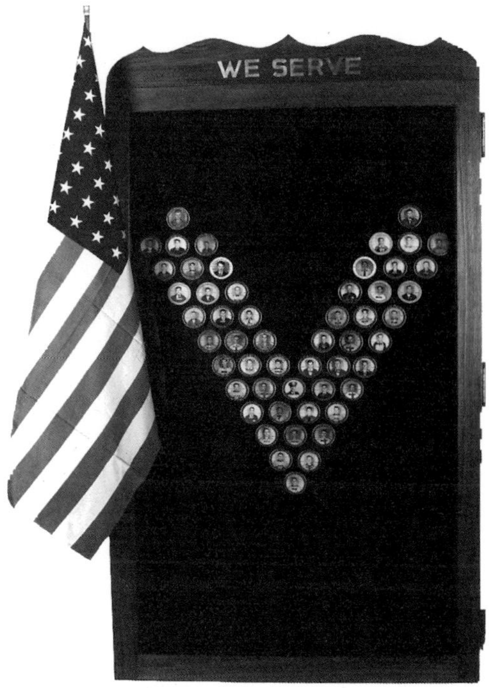

These company photo identification badges of members in the Armed Forces were used to display a "V" for victory in 1942.

The war's uncertainties bothered Irl, and he regularly reminded the membership that the war could be lost unless everyone worked together to support the United States and its allies. Irl personally corresponded with scores of servicemen and women who were Woodward members and, in December 1942, he began the "Dear Gang" column in the *Prime Mover Control* expressly to inform servicemen and women about what was going on at the company. In fact, during the war there was very little printed in the *Prime Mover Control* that was not service-related. All service personnel were promised when they left that their jobs would be held for them, and Irl was generous with company money in paying them their bonuses and giving them gifts. When he traveled on behalf of the company, he made it a point to stop and see Woodward service personnel who were stationed near his destinations.[36]

This close contact paid off at least twice when the company was able to make improved products that saved servicemen's lives. In the first case, Woodward serviceman Joe Terranova had returned to Rockford on leave and informed Forrie Drake that the Signal Corps communication system was defective. Forrie soon perfected a two-way constant communications reel that was adopted by the Signal Corps as an important improvement. No compensation was asked for or given when, after building and donating thirty-seven of these to the armed forces,

Navy machinists receive diesel governor operating and maintenance instruction from Woodward engineer Rex Hunt, in 1942. (Behind Hunt is a Hamilton Standard propeller, also used for classroom instruction.)

Navy Rear Admiral W. C. Watts presents the Army-Navy "E" flag (for excellence in production) to Roy Estell. Irl Martin is at the left, with papers in hand.

Woodward turned the drawings over to another company which kept the services supplied. In another case, a fighter pilot by the name of Cam Perks wrote that the fighter airplane governor had a fault. When a pilot pulled out of a steep dive, the propeller became uncontrollable. It didn't take long for the company engineers to figure out how to fix the problem, but the company then proposed to the government that, because of the emergency, they wanted to temporarily work twenty-four hours per day (three shifts, eight hours, seven days a week) to make the improved governors. This violated the federal rule against working seven days a week, and the government wouldn't make an exception in spite of the war emergency. Woodward members circumvented this, as they had in the 1930s, by working seven days a week, but recording only forty hours of their time. The extra hours never showed on the books, or in the members' paychecks, but the improved governor was shipped in much higher numbers than would have been possible had Woodward stuck to government rules.[37]

By 1942, the company was prepared for the next step in the development of governors. Amos and Elmer Woodward had possessed the mechanical ability and ingenuity to come up with an idea, make a working model, and then, by means of practical experience and adaptations, perfect the working model through field testing and correction. The method was almost entirely practical and little higher mathematics was needed. By the early 1940s, however, Woodward's innovative engineering group had laid the foundation for a dramatic change in the field of automatic control. Implementing this change would require more sophisticated mathematical analysis of systems and hardware. As a result, Dr. Rufus Oldenburger was hired as a research mathematician, and the mathematical analysis of control systems began. Concurrently, the government sponsored a group of scientists at the Massachusetts Institute of Technology to study artillery fire control. They also used mathematical simulation and analysis of automatic control systems. After the war, some of the results of their work were declassified and published, and Woodward engineers discovered how "nearly parallel were their and our efforts." This was the beginning of servomechanism theory, basic to an automatic control analysis. Thus, Woodward was in on the ground floor in using what was recognized after twenty years as the proper approach to the design of any automatic control device.[38] Other major events that directly affected the company in 1942 included the hiring of a company physician, Dr. Maurice Rogers, on June 2, and the all-out effort to hire women. By June 1, the company had hired at least one hundred women and was in the process of hiring approximately twenty-five additional women each week. Preference was given to wives and female relatives of male members, and a spokesman thought the company would have over 300 female members by autumn.[39] This number was easily met as Woodward's total work force more than doubled to almost 1,600 by the end of the year.

Two major company events occurred on October 9, 1942, the dedication of the new plant and the memorial of Elmer Woodward on the eightieth anniversary of his birth. At the plant dedication, the company was presented with the Army-Navy "E" (for excellence in production) flag. Irl had known that the company was eligible for the award but had not pushed for it until he was requested to do so at a plant meeting. The Army and Navy were happy to oblige. The memorial to Elmer Woodward consisted of a shrine that housed the bell from the N. C. Thompson Reaper Works where Amos Woodward invented his first governor. Cy Erickson, the company's oldest member, dedicated the memorial.[40]

The company lost a significant member that year when George Sorensen, who had served the company both as vice president and treasurer, died. Irl eulogized him as "a typical Woodward man—no one personified Woodward principles better than he," and he told affectionate anecdotes about George's inherent honesty and personal efforts to help Irl when he joined the company. Irl also noted that George was his kind of man: "We used to argue all morning, have lunch together, and then argue all afternoon." But once a policy was decided upon, differences were forgotten and everyone pulled together.[41]

Teamwork was often a favorite topic of Irl's when he spoke before the membership. Once he even told them that they could call him any name they wanted; so long as they thought it was in the best interests of the company, they would not get in trouble. This was quite a challenge, and it would take a confident, brash soul to risk personal criticism of Irl that might not be construed as being in the "best interests of the company." Irl emphasized how open-minded he was by noting that he had been an isolationist before the war but had changed his mind once the United States had entered the war. He concluded a meeting of November 2, 1942, just before the off-year national elections, by noting, "If anybody told me how to vote I'd tell him to go jump in the lake. I'll vote the way I want to. My wife will vote the way I

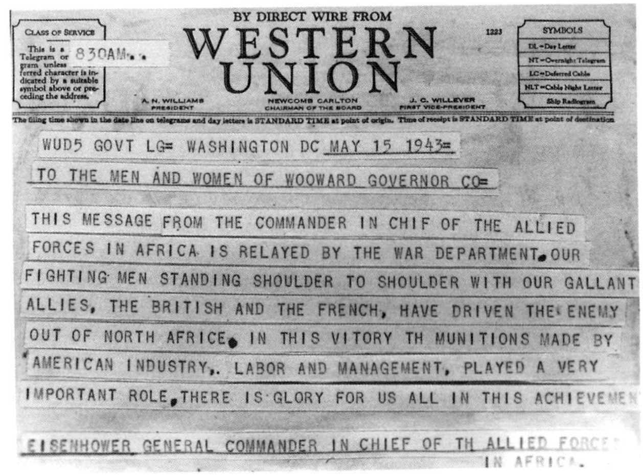
BY DIRECT WIRE FROM
WESTERN UNION

WUD5 GOVT LG= WASHINGTON DC MAY 15 1943=

TO THE MEN AND WOMEN OF WOOWARD GOVERNOR CO=

THIS MESSAGE FROM THE COMMANDER IN CHIF OF THE ALLIED FORCES IN AFRICA IS RELAYED BY THE WAR DEPARTMENT. OUR FIGHTING MEN STANDING SHOULDER TO SHOULDER WITH OUR GALLANT ALLIES, THE BRITISH AND THE FRENCH, HAVE DRIVEN THE ENEMY OUT OF NORTH AFRICE. IN THIS VITORY TH MUNITIONS MADE BY AMERICAN INDUSTRY,. LABOR AND MANAGEMENT, PLAYED A VERY IMPORTANT ROLE, THERE IS GLORY FOR US ALL IN THIS ACHIEVEMEN

EISENHOWER GENERAL COMMANDER IN CHIEF OF TH ALLIED FORCE IN AFRICA.

A telegram from General Dwight D. Eisenhower recognizing Woodward's contribution to the war effort.

The date for the ceremony described in this invitation coincides with Elmer Woodward's date of birth, October 9.

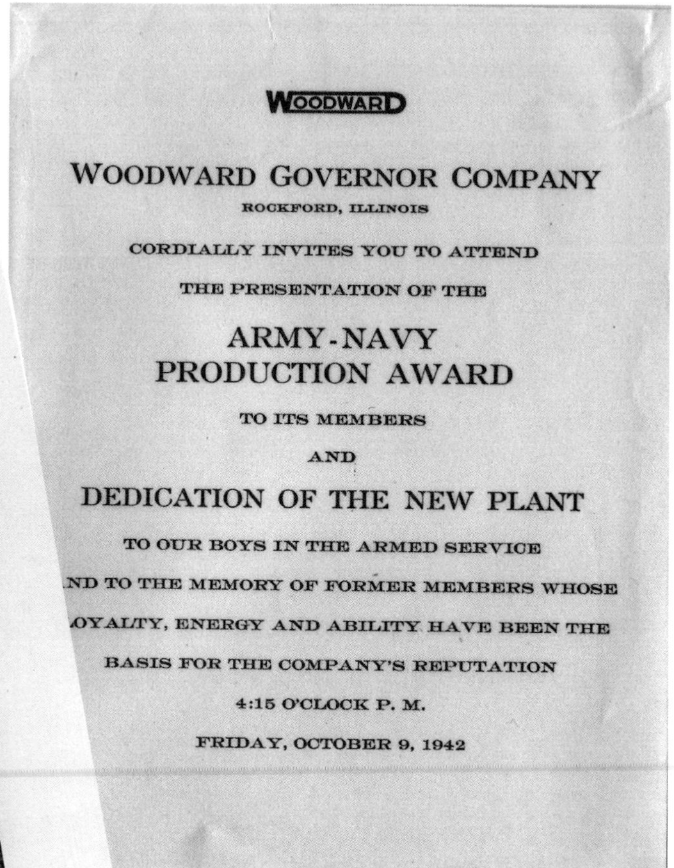

WOODWARD

WOODWARD GOVERNOR COMPANY

ROCKFORD, ILLINOIS

CORDIALLY INVITES YOU TO ATTEND

THE PRESENTATION OF THE

ARMY-NAVY
PRODUCTION AWARD

TO ITS MEMBERS

AND

DEDICATION OF THE NEW PLANT

TO OUR BOYS IN THE ARMED SERVICE

ND TO THE MEMORY OF FORMER MEMBERS WHOSE

OYALTY, ENERGY AND ABILITY HAVE BEEN THE

BASIS FOR THE COMPANY'S REPUTATION

4:15 O'CLOCK P. M.

FRIDAY, OCTOBER 9, 1942

ELMER ELLSWORTH WOODWARD

(1862 - 1940)

Elmer Woodward memorial bell, bearing the following inscription:

THIS MEMORIAL WAS DEDICATED ON OCTOBER 9, 1942, TO THE MEMORY OF ELMER E. WOODWARD BY THE ENTIRE MEMBERSHIP OF THE WOODWARD GOVERNOR COMPANY FOR THE WISDOM, PATIENCE, SINCERITY, AND HUMILITY WITH WHICH HE FATHERED OUR MUTUAL WELFARE FOR SO MANY YEARS.

IN 1868 THIS BELL WAS LOCATED IN THE TOWER OF N. C. THOMPSON'S REAPER WORKS, IN WHICH AMOS WOODWARD INVENTED THE ORIGINAL GOVERNOR, AND WHICH LATER SERVED AS THE HOME OF THE WOODWARD GOVERNOR COMPANY FOR MANY YEARS. DURING THESE YEARS THIS BELL WAS USED AS THE TIME SIGNAL CALLING AMOS AND HIS SON ELMER WOODWARD AND THEIR FELLOW MEMBERS TO THEIR LABORS. IT ALSO SERVED THE COMMUNITY AS A FIRE OR WARNING SIGNAL.

Once again, by means of the plant loud speaker system, the bell calls Woodward members to their labors and gives all time signals.

The Elmer Woodward memorial bell. The photo and text first appeared in the December 1942 issue of Prime Mover Control.

want to also."[42] This was meant to be funny and the group laughed. Not too many years later his "Dear Gang" letters were full of political advice and appeals for votes for specific conservative candidates. By the end of the year, Irl thanked the members for their herculean efforts in turning out larger production numbers each month. Just as he had given them hell whether they needed it or not, he was lavish with his praise.[43]

In 1943, sales and personnel expanded again, and gross sales, which had bottomed out at $64,000 in 1933 and had reached the one million dollar mark for the first time in 1939, were more than $17 million. This was the World War II high, and a figure the company would not surpass for a number of years after the war. Personnel also reached a World War II high. Over 1,800 men and women worked for the company that year. During this and every World War II year, the company's profits were subject to renegotiation under the War Profits Control Act, and the ultimate effect was always to sharply reduce net profits.

During this year the company gave the Curtiss-Wright Corporation, Propeller Division, a non-exclusive license to manufacture Woodward propeller governors in order to aid the war effort. Also of significance was the Board of Directors charge given to Irl Martin to have an outside firm investigate the need for a pension plan to be sponsored by Woodward.[44] This latter idea would lead to Woodward's first pension plan.

Irl continued addressing the members in plant meetings and inadvertently provided howls of laughter when he said in one meeting, "Let's work out a New Deal, pardon me, a Square Deal."[45] He also chastised the group at one time because only about one-third of them used the sun room. He wondered if it was because of their concern for modesty. He noted that people on French beaches did not wear anything, but he quickly added, "I'm not advocating it here."[46]

The frivolity of comments like these belies the seriousness and tension of the times. World War II was more than a conflict between democracy and its enemies. There was also a race for ideas, technical as well as political, and Woodward found itself right in the middle of the technical race. Sometime in 1943, the company began work on a secret government project that would last until the war's end.[47] The project was to develop a governor that would use kerosene rather than the hydraulic fluids then in use. Forrie Drake was put in charge of the engineering, and neither he nor the others he assembled to work with him—Bill Whitehead, Urban Floor, Ned Garst, and a few others—had any idea what they were building. But it had to be important, because Drake had returned from a trip to Wright Field in Ohio, where the Army Air Corps conducted much of its top-secret research, with a valise full of secret plans.

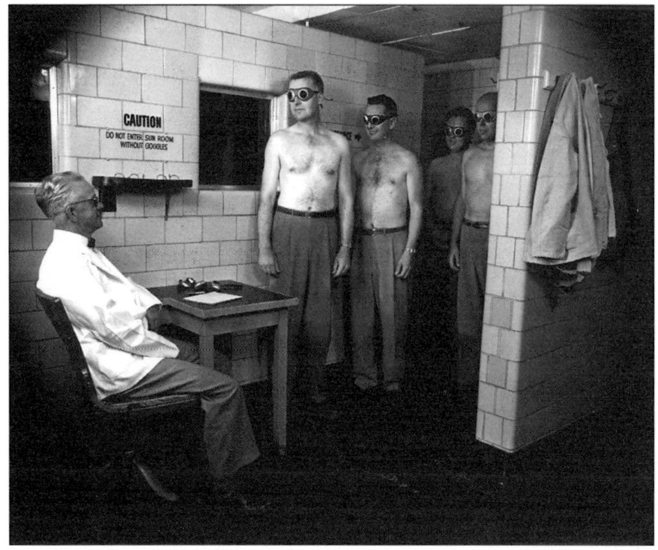

The new plant's health facilities also included a sun room which required that members wear goggles to protect their eyes from ultra-violet light.

Drake was the right man to lead the project. As head of engineering, he had put together a staff of dedicated professionals whose credentials were second to none. He was an intense, thoroughly organized man, with a sharp eye for detail and minutiae. He kept scrupulous records of every test and experiment in numbered books. His dignity and bearing had won him everyone's respect. But standing above all this was a management style unusual even for Woodward's liberalized standards. It was a style tailor-made

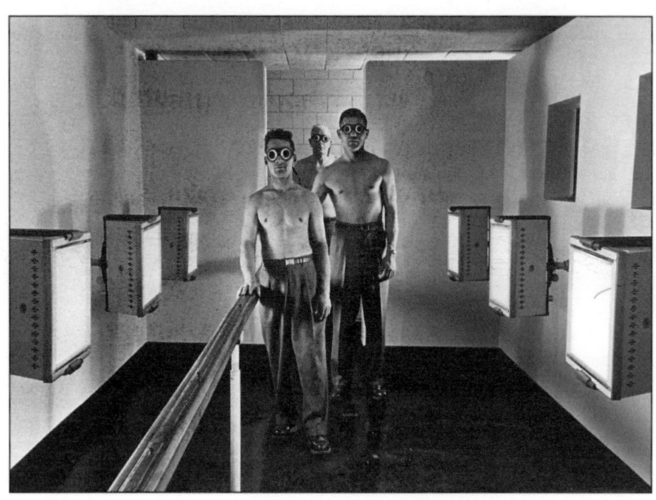

This 1953 photo shows the motor-driven belt with finger buttons that indicated the proper speed of movement through the room for maximum safe exposure.

for engineers who were talented and creative, at times eccentric, and who needed the freedom to pursue experimental leads that might not prove to be of any practical value. Drake seemed to understand that engineering was an art, perhaps the art of converting scientific principles and theories into working artifacts, and that dead ends were an inevitable by-product of experiment and exploration. It was Drake, then, who let his "wild people" follow their creative bent toward practical results, and who protected them from outside meddling.[48]

As soon as Drake returned to Rockford from Wright Field, work began on this secret, high priority project. They began by running tests on a standard double capacity airplane governor to determine pump capacity and the effect kerosene would have on the bearings. Jokes and guesses were made about this secret project, but within a month it was clear to the engineers that the project was a governor for a gas turbine—a jet engine.[49] The need was urgent, because American intelligence had learned that jet engine development was a high priority with Germany as well.

Aerial view of the North Second Street plant, about 1943. Four tennis courts were located behind the plant.

Experiments continued throughout the war. During this time, Woodward joined with General Electric and Westinghouse to conduct experiments. Test flights at Muroc (now Edwards Air Force Base in California) were held and constant adjustments were made on the experimental governors. To keep up with the work load, Woodward's development group was split into a day and a night shift so tests could be run constantly. Once, Bill Whitehead and Urban Floor worked steadily from a Tuesday morning to a Thursday afternoon, fifty-six hours without sleep, to fix a leak in a casting. Urban then went home to sleep, but Whitehead immediately took the casting to Wright Field for further tests.[50]

During the middle of this project Ned Garst, one of the development engineers, was drafted, although he was desperately needed in this project. The company tried to get him released or assigned to this project as his Army duty, but, because the project was secret, the company couldn't tell the Army exactly why it needed him. Even though the Army officer in charge requested this assignment, bureaucracy and red tape prevented his staying with Woodward even in uniform. This only made the project more difficult to carry out. Other difficulties were reflected upon by Forrie Drake, who noted:

During this period the engine limitations were being learned and new requirements set up at a rate which made the current production model obsolete in view of the one being designed which was itself inadequate to meet the requirements outlined in that morning's mail... One thing soon became clear; in order to get the most out of this new engine, its users wished to operate it just below the speed at which it would disintegrate and the temperature at which it would melt.[51]

Drake and the Woodward engineers were frustrated by what they perceived to be the opinion of some Army personnel that the Army was doing the company a favor by allowing them to stay in business and provide governors for this secret project. An experimental jet, the Bell P59, was used to test various types of governors, but the United States would have no combat jets ready until after the war ended.[52] In the interim between the defeats of Nazi Germany and Japan, Irl appealed to Woodward servicemen in Europe to send the company a governor from a turbine propelled German plane, but it is unlikely that this appeal ever succeeded.

From the company's point of view, the main business advantage of this secret project would be the ties developed with other companies, especially General Electric. After the war, these ties would be further strengthened when Woodward began producing governors for G.E. gas turbine engines. Since then, G.E. has become Woodward's biggest customer.

During the last two years of the war, Irl and other company leaders were preoccupied with production so that Woodward's part in winning the war was not hindered by sloppy workmanship, delays, absenteeism, or conflicts within the plant. Irl constantly preached harmony and consensus at his plant meetings and in his "Dear Gang" letters to servicemen and other company members. Concern for production was underlined in early 1944, when Irl noted that more airplane governors were needed. However, war manpower requirements had risen to prepare for the invasion of Europe and to continue the successful island-hopping campaigns in the Pacific. More men had been drafted, creating a shortage of people available for work. The draft, naturally, dipped into the Woodward membership, and during the last year-and-a-half of the war the number of company members dropped significantly. Irl appealed to the remaining membership for help in finding more workers, transporting them to the plant, and keeping them happy

Lawrence "Doc" Varney passing out vitamins in 1943. This was part of the company's early programs to promote member health.

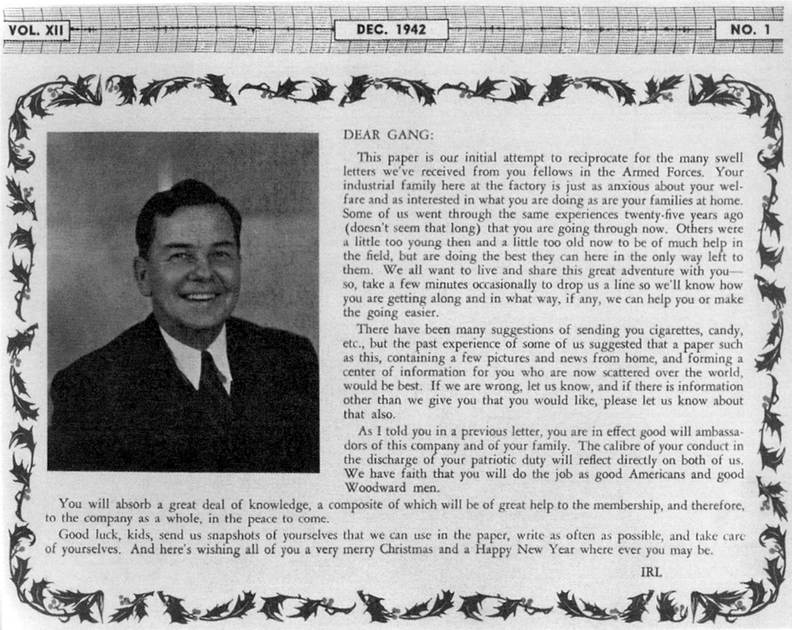

VOL. XII | DEC. 1942 | NO. 1

DEAR GANG:

This paper is our initial attempt to reciprocate for the many swell letters we've received from you fellows in the Armed Forces. Your industrial family here at the factory is just as anxious about your welfare and as interested in what you are doing as are your families at home. Some of us went through the same experiences twenty-five years ago (doesn't seem that long) that you are going through now. Others were a little too young then and a little too old now to be of much help in the field, but are doing the best they can here in the only way left to them. We all want to live and share this great adventure with you—so, take a few minutes occasionally to drop us a line so we'll know how you are getting along and in what way, if any, we can help you or make the going easier.

There have been many suggestions of sending you cigarettes, candy, etc., but the past experience of some of us suggested that a paper such as this, containing a few pictures and news from home, and forming a center of information for you who are now scattered over the world, would be best. If we are wrong, let us know, and if there is information other than we give you that you would like, please let us know about that also.

As I told you in a previous letter, you are in effect good will ambassadors of this company and of your family. The calibre of your conduct in the discharge of your patriotic duty will reflect directly on both of us. We have faith that you will do the job as good Americans and good Woodward men.

You will absorb a great deal of knowledge, a composite of which will be of great help to the membership, and therefore, to the company as a whole, in the peace to come.

Good luck, kids, send us snapshots of yourselves that we can use in the paper, write as often as possible, and take care of yourselves. And here's wishing all of you a very merry Christmas and a Happy New Year where ever you may be.

IRL

Irl Martin's first "Dear Gang" letter, from the December 1942 issue of Prime Mover Control.

once they were hired. The company was in a pinch; instead of working eight hours, many people were working ten or twelve. Irl was proud of their productivity and spurred them on by noting that if the company faltered the government might take over the plant. This was a worst case scenario, but it terrified Irl to think that a company as self-sufficient as Woodward could be commandeered by the government.[53]

The situation became so critical by mid-April 1944, that a questionnaire was distributed asking members for recommendations to make the plant more efficient. Irl discussed using the dentist chair as a barber chair, because with long working hours it was difficult to get a haircut. He also talked about adding an in-plant nursery for the children of working women. He was not anxious to do either of these, but if the situation required it, he would. He railed against the government for putting limitations on the number of hours women and men could legally work, despite the shortage of workers and the requirements for victory. Ultimately, the membership voted for longer hours than the law allowed after Irl had dramatically offered to go to jail if the membership wanted him to get the laws changed.[54]

On the eve of the June 6 D-Day invasion, Irl wrote in "Dear Gang" that the company had licensed Nash-Kelvinator to build its airplane governors after the government, for security reasons, had requested at least two suppliers. Likewise, after the Curtiss-Wright Corporation had contacted Woodward about inadvertently violating one of Woodward's patents, Curtiss-Wright was given a free license to continue to build the governor until six months after the war ended. Winning the war was more important than getting into legal battles with other companies over patent rights.[55] The company also discovered that it had given out one free license without even knowing it. Management was amused and flattered. A wrecked Japanese fighter had been scavenged and its governor sent to Woodward. It was found to be an exact copy of a Woodward governor, complete with Woodward part numbers on all the major components.

With the successful invasion of Europe, Americans began to conclude that the war was not endless and began preparations for the post-

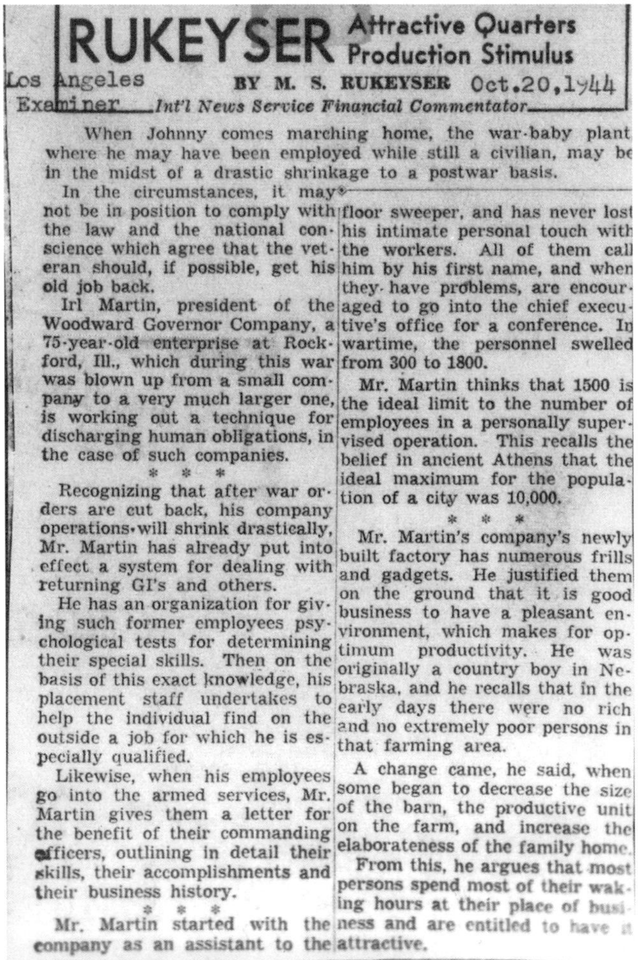

RUKEYSER Attractive Quarters Production Stimulus

Los Angeles Examiner

BY M. S. RUKEYSER Oct.20,1944

Int'l News Service Financial Commentator

When Johnny comes marching home, the war-baby plant where he may have been employed while still a civilian, may be in the midst of a drastic shrinkage to a postwar basis.

In the circumstances, it may not be in position to comply with the law and the national conscience which agree that the veteran should, if possible, get his old job back.

Irl Martin, president of the Woodward Governor Company, a 75-year-old enterprise at Rockford, Ill., which during this war was blown up from a small company to a very much larger one, is working out a technique for discharging human obligations, in the case of such companies.

* * *

Recognizing that after war orders are cut back, his company operations will shrink drastically, Mr. Martin has already put into effect a system for dealing with returning GI's and others.

He has an organization for giving such former employees psychological tests for determining their special skills. Then on the basis of this exact knowledge, his placement staff undertakes to help the individual find on the outside a job for which he is especially qualified.

Likewise, when his employees go into the armed services, Mr. Martin gives them a letter for the benefit of their commanding officers, outlining in detail their skills, their accomplishments and their business history.

* * *

Mr. Martin started with the company as an assistant to the floor sweeper, and has never lost his intimate personal touch with the workers. All of them call him by his first name, and when they have problems, are encouraged to go into the chief executive's office for a conference. In wartime, the personnel swelled from 300 to 1800.

Mr. Martin thinks that 1500 is the ideal limit to the number of employees in a personally supervised operation. This recalls the belief in ancient Athens that the ideal maximum for the population of a city was 10,000.

* * *

Mr. Martin's company's newly built factory has numerous frills and gadgets. He justified them on the ground that it is good business to have a pleasant environment, which makes for optimum productivity. He was originally a country boy in Nebraska, and he recalls that in the early days there were no rich and no extremely poor persons in that farming area.

A change came, he said, when some began to decrease the size of the barn, the productive unit on the farm, and increase the elaborateness of the family home.

From this, he argues that most persons spend most of their waking hours at their place of business and are entitled to have it attractive.

war. Woodward prepared for the future when it hired the Psychological Corporation to again test the members' aptitudes and potentialities. This especially included those members serving with the armed forces and who were home on leave. The testing began in July 1944, and Irl thought an accurate knowledge of a person's aptitudes would aid both the company and the individual. In a "Dear Gang" letter for that month he wrote: "…the solution to the postwar problem is not just a matter of jobs, but of the accurate determination of the type of endeavor for which the particular individual is best fitted… Full utilization of our best potentialities will produce an inner satisfaction that will make money a by-product rather than a goal. That is what we want, isn't it?"[56] The answer to the question depended on the individual, but many people—including Irl Martin—feared that once the war was over the country would not be able to absorb all the newly unleashed manpower and that a depression would result. The experts were wrong in part because they underestimated the demand for consumer goods and services.

At Woodward, all members in the military were considered to be on extended leaves of absence. In lieu of life insurance, which the carrier cancelled, the company sent all members in the service $100 to buy National Service Insurance. Bonus money was continued for the year they entered the service, and seniority was based on the number of years with the company plus service time.[57] These were

This column, written by syndicated columnist M. S. Rukeyser in 1944, brought national recognition to Irl Martin's personnel policies. Rukeyser (or the paper's typesetter) made one mistake, however, in the first full paragraph of the second column: the number 1500 was incorrectly mentioned as the ideal number of employees for a company. Irl believed that 500 was the maximum ideal number.

Woodward members mustering out of the military in 1945, shown here taking vocational tests in the North Second Street plant auditorium. Test results were used to assist returning veterans in career decisions. Photos of company members who served in the military during World War II line the walls.

farseeing policies and promises, particularly since the company expected major dislocations once the war was over, but they were similar to the company's other attempts to let the services know the talents of each Woodward member who joined. Irl gave each draftee and enlistee a letter for his commanding officer which listed the skills, accomplishments, and business history of the individual.[58]

The war was being won by early 1945. The Germans' attempted breakout had failed in the Battle of the Bulge, and Soviet armies were rapidly heading toward Berlin from the East. In the Far East, Japan was isolated, hungry, nearly without oil, and in general retreat.

At the company, membership continued to go down as Irl planned for the postwar corporation. He had developed his vision on that subject throughout the war and had given hints about his ideas in plant meetings and his "Dear Gang" letters. He would attempt to use Elmer's example, his own ideas, and information he received from company members to perfect a corporate structure still germinating in his mind.

The Formation of the Corporate Partnership

1945-1954

Irl had expended considerable energy during the war to prepare the company for the post-war period. He could not predict what would happen in the country—no one could—but he knew for certain what he wanted for the company, and here his foresight was remarkable.

Riding the crest of his vision was the question: what to do about company members? There were two groups. In the first were those who had joined the company before or during the war, had enlisted or been drafted, and who had been promised their jobs and seniority when they came back. In the second group were those hired to take their places. Many were women; all had been told that their jobs would end when the war ended.[1] But Irl felt a responsibility to them as much as to the returning veterans. He simply would not turn the war hirees out into the streets.

When V-J Day was announced in August, Woodward closed for the rest of the week, and Irl used the time off to iron out his personnel plans. The next week, representatives from other local companies were invited into the Woodward auditorium to interview the departing war hirees. Based on their skills and work records, they were granted interviews with those companies that had slots open for them. When the interviews were over, those who wanted a job found one with a good company.[2] This successful effort illustrated a typical trait of Woodward: it rewarded loyalty and service, and it was compassionate even toward those whose services were temporary. The *Rockford Morning Star* enthusiastically endorsed this action as "a model for war plants everywhere."[3]

The Production Control records area, located on the mezzanine overlooking the machine floor in 1945, where the progress of every production order in the shop was followed.

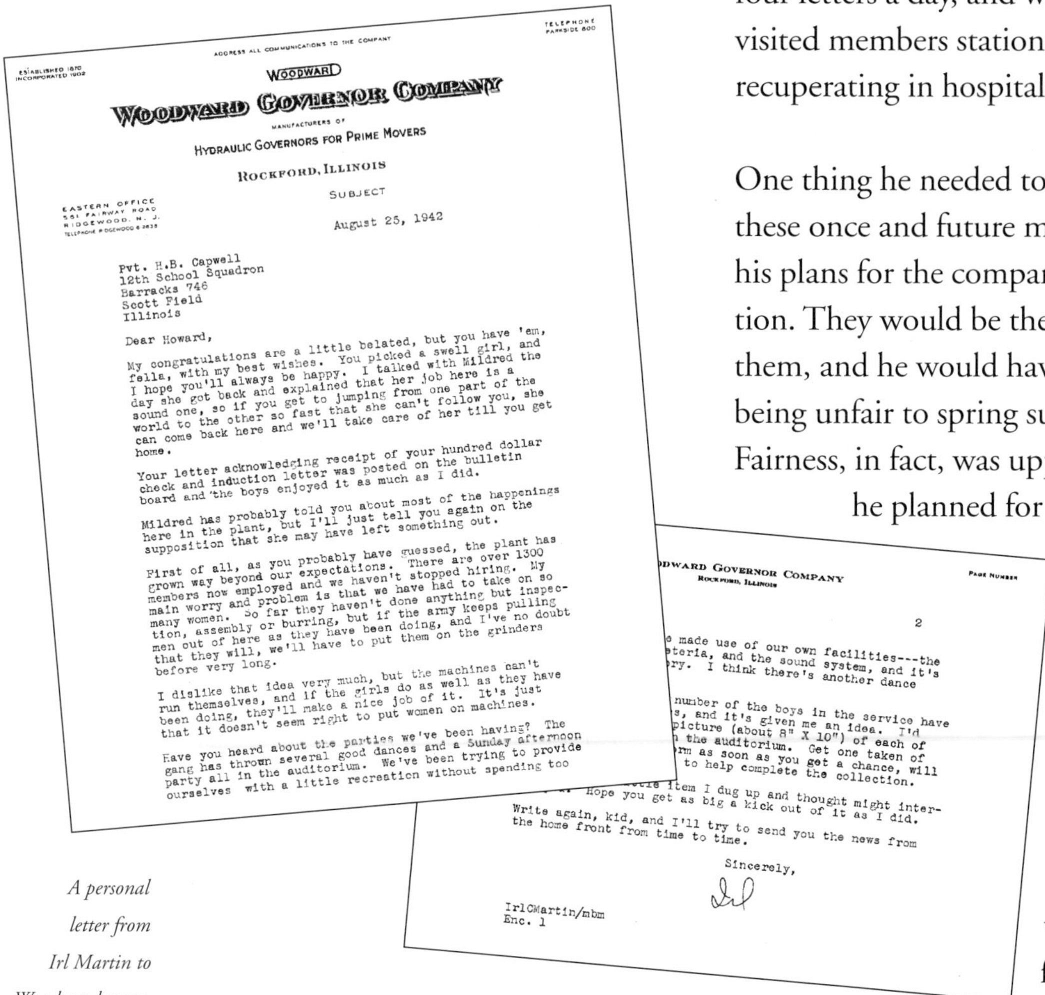

ADDRESS ALL COMMUNICATIONS TO THE COMPANY

ESTABLISHED 1870
INCORPORATED 1902

TELEPHONE
PARKSIDE 800

WOODWARD

WOODWARD GOVERNOR COMPANY

MANUFACTURERS OF
HYDRAULIC GOVERNORS FOR PRIME MOVERS

ROCKFORD, ILLINOIS

SUBJECT

EASTERN OFFICE
561 FAIRWAY ROAD
RIDGEWOOD, N. J.
TELEPHONE RIDGEWOOD 6-2635

August 25, 1942

Pvt. H.B. Capwell
12th School Squadron
Barracks 746
Scott Field
Illinois

Dear Howard,

My congratulations are a little belated, but you have 'em, fella, with my best wishes. You picked a swell girl, and I hope you'll always be happy. I talked with Mildred the day she got back and explained that her job here is a sound one, so if you get to jumping from one part of the world to the other so fast that she can't follow you, she can come back here and we'll take care of her till you get home.

Your letter acknowledging receipt of your hundred dollar check and induction letter was posted on the bulletin board and the boys enjoyed it as much as I did.

Mildred has probably told you about most of the happenings here in the plant, but I'll just tell you again on the supposition that she may have left something out.

First of all, as you probably have guessed, the plant has grown way beyond our expectations. There are over 1300 members now employed and we haven't stopped hiring. My main worry and problem is that we have had to take on so many women. So far they haven't done anything but inspection, assembly or burring, but if the army keeps pulling men out of here as they have been doing, and I've no doubt that they will, we'll have to put them on the grinders before very long.

I dislike that idea very much, but the machines can't run themselves, and if the girls do as well as they have been doing, they'll make a nice job of it. It's just that it doesn't seem right to put women on machines.

Have you heard about the parties we've been having? The gang has thrown several good dances and a Sunday afternoon party all in the auditorium. We've been trying to provide ourselves with a little recreation without spending too

DWARD GOVERNOR COMPANY
ROCKFORD, ILLINOIS

PAGE NUMBER

2

e made use of our own facilities---the
eteria, and the sound system, and it's
ry. I think there's another dance

number of the boys in the service have
s, and it's given me an idea. I'd
picture (about 8" X 10") of each of
n the auditorium. Get one taken of
rm as soon as you get a chance, will
to help complete the collection.

item I dug up and thought might inter-
Hope you get as big a kick out of it as I did.

Write again, kid, and I'll try to send you the news from the home front from time to time.

Sincerely,

Irl

IrlCMartin/mbm
Enc. 1

A personal letter from Irl Martin to Woodward member Howard Capwell, serving in the Armed Forces during World War II. This is just one of hundreds of letters Irl wrote to Woodward members serving at home and overseas during the war.

The company had little trouble rehiring those veterans who wanted to return. Throughout the war the company had made a determined effort to stay in touch with as many of them as possible. *Prime Mover Control* was published twice a month, and most of it was devoted to stories and pictures about Woodward servicemen and women, rather than products, conferences, and other company events. The company also sent them Christmas presents, and members in Rockford besieged them with cards and letters. Irl himself averaged about four letters a day, and when he traveled he always visited members stationed near his destination or recuperating in hospitals.[4]

One thing he needed to know were the opinions these once and future members might have about his plans for the company's postwar reorganization. They would be the people most affected by them, and he would have regarded himself as being unfair to spring surprise changes on them. Fairness, in fact, was uppermost in Irl's mind as he planned for the years to come after the war. Nor was it lost on members stationed all over the world that their company was thinking about them and their future. They responded by providing Irl with a wide array of views that helped him fine-tune the Corporate Partnership ideas he was nursing during the war. And Irl himself built up a reservoir of good will and confidence among the membership that would make the transition to Corporate Partnership easier.

As Irl's ideas evolved during 1945, President Harry Truman was learning about the existence of the Manhattan Project to build the atomic bomb. He was also trying to get along with the Russians and British at the Potsdam Conference, where the squabbling over how the map of the world ought to look helped start the Cold War. And he was doing his best to convince Japan to surrender unconditionally, as Germany had in May.

After atomic bombs were dropped on Hiroshima and Nagasaki, Japan finally surrendered. What this meant to American servicemen is symbolized by the sudden turn of events on the American troop transport *General Henry Taylor*, then crossing the Pacific on its way to Japan. Abruptly, the ship's loudspeaker announced, "Now hear this! This is the captain. Watch the shadow of this ship." He paused, then continued, "as it turns toward New York." And to the wild cheering of 3,000 veterans of the European theater, the ship did just that. Also symbolic was President Truman's removal of the model gun on his desk and its replacement by a tiny plow.[5]

The United States and her allies had won the war, but winning the peace would not be so easy because Allied unity broke up once their common adversaries were defeated. Within a year, a bipolar world developed, the Soviet Union and communism on one side, the United States and the free world on the other. The division was recognized and described by Winston Churchill in his "Iron Curtain" speech delivered at Westminster College in Fulton, Missouri, on March 5, 1946.

The Cold War had begun, and it would dominate the lives of the post-World War II generation. The United States reacted to this potential danger by standing up to what the country's leaders determined were aggressive designs on the part of the Soviet Union and its allies. Both sides squandered the good will that had accrued during World War II. A collision appeared imminent by the late 1940s.

Europe and Asia were in ruins, and the United States was the only power that came out of the war physically unscathed. Foreign policy dominated the immediate post-World War II years as the U.S. struggled to promote a stable Western Europe and Asia friendly to American interests. America helped to establish the United Nations in 1945. It agreed on peace treaties with the Axis satellites in 1946, promoted the Truman Doctrine and Marshall Plan in 1947, supplied West Berlin during the Russian blockade in 1948, and set up NATO in 1949. Troops were sent to aid South Korea in 1950, where they remained after the truce of 1953.

Engineers (from left): Norm Noling, Forrie Drake, Ritchie Dewey, and Ernie Swarts, get their heads together over a new design in 1945.

The next year, the French were defeated at Dien Bien Phu in Indochina and, shortly afterwards, the United States began the process of replacing the French and supporting a new state called South Vietnam.

The world was becoming a dangerous place, even with no major wars in the immediate offing. The United States was the only country with atomic weapons from 1945 to 1949 when the Soviet

The first company airplane, a 1941 Culver Cadet, was purchased in 1945 and converted from a fixed pitch prop to a constant speed propeller. At Machesney Airport are Woodward members (from left) Del Zeigler, Mac Garard, and Louis Bertchie.

Union exploded its first atomic bomb. The bombs became bigger and the world became scarier after November 1952, when the United States exploded the first hydrogen bomb, and Russia followed suit nine months later.

None of this was foreseen in 1945 when people of good will had hoped that the "Good War" that had toppled Germany and Japan would be the last war necessary. They were partly correct: the nuclear powers avoided war among themselves. But each side supported smaller conflicts among client states that served to heighten the Cold War.

Domestically, the United States emerged from World War II with some economic and political doomsayers predicting a return to the Depression. American industry had become the "Arsenal of Democracy" and, by anybody's standards, it was a force of unprecedented size and power. There was, however, a general fear that once the war was over industry would shrink back to its pre-war size. To most people the decade-long Depression was still fresh in their minds, with the effect that they overlooked the massive savings of the war years when many consumer goods, like automobiles, washers, stoves, and refrigerators, were unavailable. In addition, the average weekly wage had risen from $24.20 to $44.39 since Pearl Harbor and, with their large amounts of overtime, people had saved $136.4 billion in bank accounts, Liberty Bonds, and liquid securities. Moreover, the Roosevelt administration had started social security, the G.I. Bill of Rights, and 52-20 unemployment compensation, which guaranteed returning veterans fifty-two weeks of unemployment compensation at $20 a week. With the government campaign to get women war workers to return home, and the marriage and baby boom that followed the war, plenty of jobs were available and a consumer revolution began as demand outpaced supplies. Although there was a brief recession in the late 1940s, the country left the Depression far behind.

In Rockford, when the veterans returned, they found their companies converting from war-related necessities to consumer-oriented products. Automobiles were made again and the desire for personal transportation and other consumer products burned holes in the pockets of many a veteran. Once a car was acquired, fox or squirrel tails were tied to radio antennas, and before long baby booties were hung from rear view mirrors. Money became "moola," and people greeted each other with the phrase "long time no see."[6] Life was far from perfect though. The cost of living shot up once wartime price controls were lifted, and strikes, almost nonexistent during the war, returned in earnest in 1946. When two railroad unions struck that year, President Truman threatened to nationalize the railroads, and he even threatened to draft all railroad men regardless of age, a radical move that did end the strike. He was just as tough in ending a coal miners' strike that threatened the nation's electricity and industrial power.

Irl Martin congratulates Walt Thorell after presenting him with a 25-year Service Pin at the 1945 Fall Party.

Meanwhile, Irl and the leaders of the Woodward Governor Company had too many people eager to work in a company whose customers could not quickly convert to peacetime products. With the war's end, Woodward's business dropped drastically. But once the company got down to an appropriate size, new orders would keep it prosperous.

Irl had tremendous confidence in his ability to lead the company to become what he hoped would be "the industrial Utopia that we would all like to see...in which every individual is perfectly happy."[7] During the latter part of the war, and through much of the winter of 1945 and the first nine months of 1946, he struggled with his emerging plan. He sought advice and criticism to fine-tune the plan before putting it into effect. His goal was to build on the firm footing established by Elmer Woodward and to remake the company into a Corporate Partnership that would be run like an industrial family. This family would require responsibilities of all its members as well as teamwork, ability, sportsmanship, and a high quality product. Because Irl was such a patriotic admirer of the United States Constitution and its Bill of Rights, he looked to those documents as guides in establishing an industrial democracy.

Like many strong business leaders with benevolent concern for their employees, Irl had an authoritarian side that was tempered by his openness to opposing views. He insisted that members be loyal to company principles. If they couldn't be, they should leave, because Irl was

During this plant meeting in 1945, Irl Martin introduced four Woodward servicemen. Photos of all Woodward members serving in the Armed Forces line the auditorium walls.

This Annual Meeting of Shareholders was held in conjunction with a plant meeting for all members on April 25, 1946.

Seated on the stage (from left) are: Walt Thorell, vice president and director; Roy Estell, vice president; Walter Woodward, director; Paul H. Davis, consultant and former director; Estelle Milne, assistant secretary; Charles Turner, assistant treasurer; Irl Martin, president and chairman; Charles Thomas, attorney; Eugene Abegg, director; Ward Ring, treasurer; Bill Taylor, director.

determined to run the company his way. But he also challenged his membership on his one fear, of being wrong. "If you catch me being wrong," he said, "I'll go down on my hands and knees," but, he added, "that hasn't happened yet."[8] He never expected it to happen either.

Between 1944 and 1946, Irl solidified and institutionalized the major company tenets into a new package. He continued to look for loyalty and support, as well as new ideas, as he tested his thoughts in plant meetings and in "Dear Gang." He encouraged businessmen to get into politics to support the benefits of business and private initiative. He urged the private business community to reconstruct the nation "on business methods, hard work, and moderation," and to "guarantee its integrity by open, above-board procedures and benefits to each in proportion to the value of his or her contribution to the total result." Irl felt strongly that "hard, honest work is the answer, not government doles and WPA."[9]

Irl was obsessed by fairness. Not only square dealing and honesty in his own personality, but fairness as a basic ethical value were tied to the example he had learned from Elmer Woodward. His feelings about this were expressed at a plant meeting in September 1945:

I'm just as honest as I was ten years ago...and just as honest as the old man... I took the reins exactly as he left them and have been operating [the company] the same way. I believe he was right then and that he is right now.

Then, pressing his point for the promotion of ideas that became the Corporate Partnership, he noted, "We can hook this damn thing together so that nobody can doubt [its fairness]." He then offered some preliminary articles of company fairness that included figures for base pay; a base rate of dividends; the amortization of buildings, machinery, and men; and insurance and profit sharing bonuses. Finally, he pleaded for understanding and support by noting, "We've got to get to the point where we're 100 percent sure of each other—you of me, me of you."[10]

Irl intensified the discussions of the Corporate Partnership plan from the fall of 1945 until it went into operation on October 1, 1946. In the interim, the evolving plan was the hottest topic in regular plant meetings and in Irl's "Dear Gang" letters. A tremendous amount of time and energy went into this effort. The end result was worth it, since the basic plan with some amendments has served the company and its stockholder and worker members very well since that time.

The basics of the 1946 plan used the principles of the partnership form of industrial organization to run the corporation, hence the term "Corporate Partnership." A true partnership could not be used because the law did not have a liability limit for partnerships, and without the limit each partner would be equally liable if the company went into debt or out of business. This was too great a risk for individual workers to take on, so the corporate form was maintained, but partnership goals were promoted.

Under the plan, the partnership existed between stockholders and workers, and workers were themselves encouraged to become stockholders. The stockholders provided the working capital, machinery, and buildings. The workers provided skills, energy, and direction. Each would be fairly remunerated from the company's total income, which would be divided according to the relative contribution of each. The return to stockholders was easier to determine since figures could be compared to the rate of dividends provided by other successful corporations. The return to worker members was more difficult to determine, but the problem was solved by letting the workers participate in determining the relative value of their co-workers. A Member Evaluation Committee, made up of a cross section of managers and supervisors, added its own evaluation to that provided by co-workers and supervisors, and arrived at a master list in each plant which ranked all members from the most valuable to the least valuable. This system was based on "man evaluation" as opposed to "job evaluation," the popular approach to wage setting. Company income beyond that needed for a prudent reserve would be used to pay the worker members profit sharing and stockholder members an extra dividend.[11] This income was divided between workers and shareholders based on a formula which measured the relative "investment" by workers compared to shareholder investment. Everyone was encouraged to live on his regular income and not to count on profit sharing. If one came (and it almost always has), it should be used for extras or as savings for future needs.

The whole plan was predicated on the idea of an industrial family, where each member received remuneration based on his relative worth. Mutual accord was promoted by the promise of fair treatment and peer review. Company success was promoted because stockholders and workers saw mutual benefits by cooperating in the purchase of facilities and the maintenance of worker efficiency. Partners, in effect, were paid a "premium for harmony and efficiency and a penalty for discord and indolence."[12]

The old Horizontal Committee system, originally a vehicle for company change and governance was gone. It was replaced in mid-1945 with the Vertical Committee system that aimed to educate members and to use their "surplus energy" to look after the interests of the corporate family.[13] The six Vertical Committees were: Legislative, Finance, Health, Recreation, Veterans, and Open Door. These committees were to play an advisory role to management. Irl remained very much in charge, but he wanted input on issues of mutual interest.

Bill Whitehead (right), Aircraft Governor sales manager, confers with A. D. Murphy of Pratt & Whitney Aircraft Division, United Aircraft Corporation, in 1947.

Each of these six committees had a specific purpose. A senior executive chaired each one, and a junior executive served as assistant chairman. They were appointed, but the rest of each committee were elected from the membership (two from supervisors, four from non-supervisors). Anyone interested in serving on a particular committee were encouraged to run. The system established a grass roots democracy within the company and helped it to flourish.

Irl's goal was to promote the sharing of information that would educate all the members and lead them to a consensus on company actions. Irl personally believed that the country faced the danger of increased government control because individuals failed to take responsibility for their own actions. The Vertical Committees would get the membership involved in everything from world and national affairs to detailed company issues, like the lunchroom menu. In the process, each member would be better educated and more willing to take responsibility in the world beyond his or her own job. Good citizenship was the goal both within and without the company. Within that framework, each committee member and all other company members were supposed to put the welfare of the company ahead of their personal welfare.

The Legislative Committee's purpose was to study proposed legislation and recommend to the corporate family the laws that should be passed or defeated on the basis of their potential effect on the company and its membership. The committee also investigated the qualifications of candidates for public office and posted its recommendations on the bulletin board just prior to elections. While no one was required to vote for the committee's choice, local politicians quickly realized the implications of potential block voting and actively sought the endorsement of this committee. The idea was to encourage independent thinking, but also to arrive at a consensus to benefit the company. Disagreements within the committee were less likely but, even if members did disagree, Irl claimed he would vote for the majority recommendation because—

if, as head of this company, I can't sell them into my way of thinking it is just as well that I accept their judgment—they may be right. Among sincere people a source of judgment generally is appreciated if its quality stands the test of time.[14]

After its first two years, Irl praised the Legislative Committee for its "judgment and civic integrity." The committee worked best and had its greatest impact at the local and regional level. A number of Woodward members had been elected to public office and the company's influence was on the rise. Irl saw this success as the company's contribution "to the political welfare of our community, state and nation."[15] However, despite his deference to this committee's judgments, Irl was quick to promote his own political views in his "Dear Gang" letters. In national election years, Irl was sure to mention "his candidate" with comments such as, "I like 'Ike' but I like Taft better."[16]

The other five committees dealt primarily with internal matters. The Finance Committee controlled the company's charitable giving and, in conjunction with the Legislative Committee, all other legislation that affected the taxes the company owed. The Health Committee did everything from advising the cafeteria on menus and prices to promoting health maintenance and accident prevention programs. The Recreation Committee and the Veterans Committee had self-evident duties, while the Open Door Committee took on any problem not specifically covered by the other committees. Within a short time after World War II, the Veterans Committee was no longer needed. Subsequently, a Tax Committee was established which took tax interests away from the Finance Committee and left it with the responsibility of charitable contributions only.

Al Gustitus makes the first strike on the new four-lane bowling alley, located near the Cafeteria, in 1947.

The committee system has been a success and has remained an influential part of the company. Its place as one of the cornerstones of the Corporate Partnership is secure.

A group of machinists pose with Ward Ring, vice president in charge of Physical Facilities, about 1953.

Kneeling (from left): Bob Eberle, Hubert Brockway, Vern Engebretson, John Lindvall, Harley Bohl, Pat Murphy, Ben Schleicher, John Hawkins, Thor Berglund. Standing (from left): Tony Ruskavage, Dick Jansen, Bob Ball, Ward Ring, Art Fairgrieves, Jack McMillan, Ken Collison, Jesse Hutter.

The Corporate Partnership itself quickly became institutionalized, and Irl's 1947 summary of the company's philosophy seems as appropriate today as it was then: "...the principal reason [Woodward] has had reasonable success and no personnel difficulties...is because it is operated as a means to a satisfactory life rather than just as a means to money making."[17] And a satisfactory life was possible because the Corporate Partnership made Woodward an unusually attractive place to work. Woodward was ahead of its time in promoting a membership system that assumed a mutually supportive relationship between management and workers. Promotion came through a series of jobs that encouraged a lifelong career and loyalty to the company. The Corporate Partnership also had built-in mechanisms that promoted idea-sharing and mutual power within and without the corporation. Moreover, once the system had begun, benefits such as medical and dental services, a pension plan, and a deferred profit-sharing (savings) plan, were added in the first decade. Also, in a highly unusual move, a chaplain was engaged to handle counseling and provide for the moral welfare of the members.

Some of the technicalities of the Corporate Partnership were tinkered with during the 1945-1954 period, but its basic structure remained intact. During this decade, the company's membership would slowly rebound from the postwar low of about 400 to almost 800—a figure 300 higher than Irl's ideal of 500. He believed that the company could not be efficiently, fairly, and intimately run under the partnership plan with many more than 500 members under one roof. One reason for the expansion to Fort Collins, Colorado, in late 1954, then, was Irl's concern that membership in the Rockford plant was becoming too large.

While membership doubled during this decade, sales also climbed from a postwar low of $3.9 million in fiscal 1946 to over $14 million in the Korean War years of 1952 and 1953, before dropping back slightly to $13.5 million in 1954.

Earnings remained strong throughout this period, and profit sharing was equally generous. Stockholder members got extra dividends and saw the book value of their stock go up annually.[18] The company continued to be run conservatively and cautiously but very successfully. Married women were laid off at the end of World War II, but many women were rehired and others recruited during the Korean War. By this time, Irl had changed his ideas on the suitability of women in non-traditional jobs.

A few months after the end of World War II, *Prime Mover Control* became a monthly publication. It reverted from personal news of the wartime members back to company news and product details. Irl's "Dear Gang" column continued. The company also began an annual report issue of the PMC in January 1947 that summarized major company events by committee or department for the preceding year.

Product development slowly modernized and changed the company between 1945 and 1954. World War II necessities had led to the expansion of the experimental laboratory, begun in the late 1930s. This lab had experimented with governors for gas turbine engines and continued after the war to seek solutions to this problem, as well as problems in the diesel and water wheel fields. Under the direction of Forrie Drake, longtime chief engineer and new vice president for Engineering, the company had developed a number of new products and improved the old ones by 1954. These were team efforts, but they were ably led by engineers Bill Whitehead, Urban Floor, Tom Larsen, Luke Avery, Gil Gillespie, and George Parker.[19]

One of the most important developments to come out of this effort was the 1307 gas turbine fuel control, a sophisticated governor designed specifically for the General Electric J79 jet engine. By late 1954, this control had won the endorsement of G.E. engineers as the control of choice for the J79. Still to be settled were price and production capability, the latter being of primary concern to G.E. As the development had progressed, so had the pricing and production studies. By late December, Norm Noling, manager of the Gas Turbine Governor Department, had prepared and presented to G.E. prices and delivery schedules as requested, a listing of production machinery and tooling available and anticipated, and a detailed account of Woodward's productive performance over the years—listing specific water wheel, diesel, turbine and aircraft governors and specific quantities and customers. Discussion of production capability continued, culminating in acceptance of Woodward's ability to meet General Electric's requirements.

The truck-mounted jet engine in this 1948 photo was used in some of the company's earliest development of a gas turbine aircraft control. To avoid the problem of engine noise in the plant's residential neighborhood, the mobile test stand was transported between the plant and this test site at Machesney Airport.

This cutaway view of a PG governor shows the type of parts found in many governors. The PG was used in pumping stations, on ships, and elsewhere.

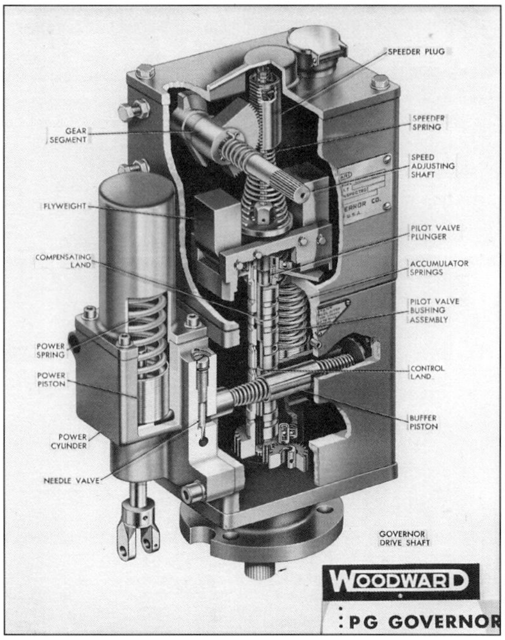

It was an especially opportune time to negotiate this contract, because sales of diesel governors and propeller governors were declining, and because plant capacity had increased with the opening that month of the plant in Fort Collins. The contract had the potential to provide about $5 million in sales per year.[20] Sales of the 1307 control did not occur immediately, but this contract began the especially close relationship that General Electric and Woodward have had since 1955. Most importantly, gas turbine engine governors would in time produce the company's largest sales volume and provide a major boost in the company's growth beyond the 1950s. Despite the potential, Irl was very dubious about the proposed agreement, but he eventually was convinced that the company should make a special effort in gas turbine development.

This IC governor was shipped to Fairbanks-Morse in 1942 to control one of their diesel engines. The IC, which was the first type of Woodward governor built for diesels, was designed in 1933 for Fairbanks-Morse.

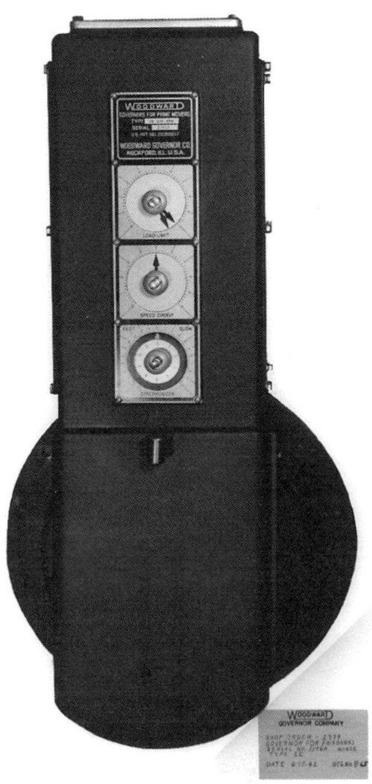

A number of other products were developed during this era. One of the most significant was the PG governor, which was perfected in late 1948 and released for sale shortly thereafter. It was built for diesel engine and steam turbine use and was an improvement over the SI governor which had been a company success since 1933. The compensating system of the PG governor both improved and simplified its predecessor. It was also designed to be interchangeable with the SI, whose owners would be encouraged to buy it. The main improvement offered by the PG was a completely new and frictionless compensating system. Consequently, the governor was "greatly improved in stability and sensitivity." By 1950, the company described the PG as superseding the SI and hinted at a "practically complete changeover" within a few months. Sales increased for the next few years for this model and its adaptations.[21]

By the end of 1950, PG governor sales to pipeline pumping stations had increased so rapidly, and diesel engine governor orders were arriving at such a fast rate, that lead times of six months became the norm. The Diesel Governor Department was on the verge of being overwhelmed because the volume of business was the greatest in the company's history—surpassing the previous records set in World War II.[22]

Orders increased again in 1951 before declining in 1952, which allowed a normal delivery period of three months or less. Steel shortages reduced the number of new locomotives, trucks, buses, oil and gas pipelines, and caused a 25 percent drop in unfilled orders. Shipments were reduced slightly, but the department's ability to quickly deliver standard governors helped keep sales up.[23]

Research led to regular improvements in locomotive governors, in diesel engine and steam turbine-driven pumping applications, and in the modification of a standard locomotive governor for use on ships equipped with adjustable blade propellers. By this time, the recently developed PG governor and the UG32 governor had largely replaced the SI and IC governors, although customers continued to buy a significant number of the older SI and IC governors.[24]

By May of 1953, the backlog of diesel orders reached its low point and orders again began increasing; however, orders in 1954 were nearly the same as in 1953. In one sense, this was a positive trend, since 1954 was a recession year, and manufacturing throughout the country declined. Governors for diesel locomotives were also in decline, and industry experts expected that they would never again reach their previous high. Nevertheless, the company had great success in adapting diesel governors for other purposes, such as in controls for adjustable pitch ship propellers, or on turbines in the chemical, petroleum pipeline, and sugar mill industries. As fiscal 1954 ended, the Diesel and Steam Governor Department was enthusiastic about its potential for the future.[25]

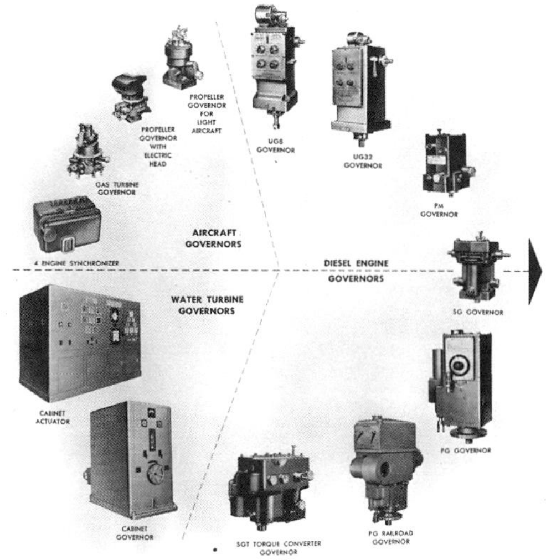

A FEW WOODWARD

PRODU

A selection of Woodward controls featured in a 1953 company publication.

During this same ten-year postwar period, the Water Wheel Governor Department expanded consistently and rapidly. During World War II, there had been little activity in the hydraulic turbine field. However, within sixteen months after the end of the war, the company had the highest number of hydraulic turbine governor orders in its history. Walt Thorell, Water Wheel Governor Department manager, predicted "many years of heavy requirements" in the field domestically and abroad.[26] His prediction proved to be accurate.

Cabinet actuators for hydro plants had greatly improved and simplified governing principles since their introduction in the 1930s. They remained state-of-the-art in the late 1940s, as changes ensured their improvement, and demand was high for new and rebuilds. As the economy blossomed in the late 1940s, an acute power shortage resulted, which

Woodward members' winter dance at Tebala Temple, downtown Rockford, in 1953.

The Christmas party for Woodward members' children, held in the company auditorium, in 1953. (We think Santa was Roy Capron.)

The Woodward family picnic was held at Macktown Forest Preserve for many years. It was an old-fashioned, bring-a-dish-to-pass gathering with hay rides, sack races, ball games, and much more.

convinced public utilities of the necessity to expand as rapidly as possible. Woodward was almost invariably chosen to provide the governing equipment for this expansion in the United States and Canada. Export shipments to other countries other than Canada declined in part due to difficulties Latin America and Europe faced in financing U.S. products. Nevertheless, smaller gateshaft governors were sold all over the world, especially to countries in South and Central America and Mexico.[27]

The boom continued during the 1948-1950 period. Although company shipments increased 78 percent in fiscal 1948, unfilled orders rose 34 percent. By the end of that year, hydroelectric units totaling over seven million horsepower were on order in the U.S. and Canada alone. Woodward governors were the governor of choice on all but a minor fraction of hydroelectric projects. The company, for instance, continued to supply additional units to the Grand Coulee Dam—a project begun in 1940 and not completed until 1950. Backlogs continued to increase in 1949 and 1950 despite the latter year's all-time record shipments.[28]

In 1950, the introduction of a new governor for smaller turbines produced still more orders, while the best seller continued to be the cabinet actuator for large hydroelectric projects. These sold well in the U.S. and Canada. Some were also sent to Brazil, India, the Philippines, and New Zealand. By this time, production capacity was filled for the next two years, and orders were being taken for delivery in 1955.[29]

The Korean War had less negative effect on sales of hydraulic turbine governors than World War II, mainly because industrial capacity could accommodate military orders without disrupting commercial interests. In addition, hydroelectric plant construction continued at a record pace.

Despite rapid expansion, the power shortage continued, which aided the sale of governors throughout the Korean War. In 1952, a new shipment record 25 percent over the old record was achieved, even though the backlog of orders increased another 5 percent. Moreover, Woodward continued to dominate the entire worldwide market for automatic and remote control of hydroelectric plants.[30]

A row of cabinet actuators being assembled under the crane bay in 1953. Standing at far right, Hydro engineer Luke Avery (in suit) confers with assembly supervisor Don Ferwerda.

During 1953 and 1954, sales and shipments of hydraulic turbine governors continued at a high but not a record level. Shipments dropped a bit in 1953, then rebounded again in 1954. Unfilled orders did just the opposite. The 8 percent drop in new orders in 1954 caused some concern because it was assumed that foreign competition in generators and transformers might soon spread to turbines and governors. Nearly all European turbine manufacturers built their own governors, but they offered little choice in governor arrangements compared to Woodward's cabinet actuators. Consequently, when the company began to receive European inquiries about specifications for the use of Woodward governors, a potential new market was seen.[31]

This Stratosphere Chamber, located in the Experimental Laboratory in 1953, duplicated the temperatures and pressures of altitudes up to 70,000 feet. Lab technician Woody Carson (right) monitors the chamber's control panel.

By the mid-1950s, the Western European countries had largely recovered from World War II and were becoming competitive with the U.S. Their recovery also meant that they would soon have enough cash to buy superior U.S. products. Woodward would begin to take advantage of this change and to consider the further advantages of manufacturing and sales facilities in Europe and/or Great Britain.

The Yard Maintenance Crew pose in front of the Elmer Woodward Memorial Bell, in 1952. Supervisor Cal Covert is in the front row, left.

Irl was concerned about the potential for foreign competition, especially of the water turbine governing equipment, since this was nearly all custom-made. He believed that overseas competitors had the advantage of lower costs and wages as well as fewer laws that restricted their business. But he remained optimistic that Woodward could meet its competition through competitive prices and superior products.[32]

Company leaders had been actively considering domestic and foreign expansion as early as 1950. In a wide-ranging interview with a writer from *Time* magazine, Irl had philosophized about the Corporate Partnership and the responsibility it put on all the company members. He also reiterated his belief that the company had remained successful because it was relatively small. People knew and cared about each other. But Irl's ideal of a maximum of 500 employees had long been surpassed, and it was clear that the increased business allowed for expansion. Irl's hope was that the company would choose a small community for a new facility because he believed that his corporate philosophy would work better in such surroundings. Irl also expected big cities to become obsolete in the future, and he predicted, "When we move again, we'll probably wind up on the western plains."[33]

In May and June of 1950, Irl and local banker Gene Abegg, who had earlier counseled company members on stock purchases, traveled to Europe "to collect business information with which to help chart the course of [the] company for a number of years to come." Abegg was a Woodward director, as well as a close friend of Irl's, and they got along well partly because their business views were similar and because they respected each other's opinions. Abegg remained a major influence on Irl for the rest of Irl's business career.

Abegg noted that Woodward was at a disadvantage in competing for European business because it did not have representation in Europe. Since there was substantial business to be gained there, he advocated that Woodward think seriously about expanding into Europe. A major obstacle in 1950 was that it was difficult for most European manufacturers to obtain enough dollars to buy American products because of the then unfavorable balance of trade. This exchange problem was seen as temporary, and Abegg believed that Woodward ought to begin preparing for the day when the European markets would open up. Then the problem would be to convince European manufacturers to use Woodward governors on their products rather than continue to produce all the parts for every machine they made. He and Irl were convinced that Woodward could make a better governor for the same money, or as good a governor for less money. Therefore, until exchange barriers lessened, Woodward would only prepare to compete in Europe, but once the barriers were gone, Woodward would be ready.[34]

While Irl and Gene were in Europe, Woodward engineer Luke Avery was in Brazil. He spent almost five months in the states of Rio de Janeiro and São Paulo coordinating governor operations between Woodward and two Brazilian organizations. Like Western Europe, Brazil lacked hard currency. It also lacked electric power to aid in the government's program of promoting industrial expansion. Consequently, there was a great potential market for Woodward to develop but, as in Western Europe, it could not develop immediately.[35] By this time, the Korean War had intervened and the company's emphasis remained domestic for the next few years.

Toward the end of the Korean War, Irl and Luke Avery took a long trip across South America to assay business prospects. Irl concluded that, as in Europe, there was a problem of balance of payments. "American dollars are chronically as scarce as amoebae are plentiful."[36] Nevertheless, Irl expected that Woodward could do better in sales to private industry than to government agencies. Private industry, in general, was more geared toward long-term economies, and Woodward with its good product, service, and financing terms could compete successfully.[37]

In the search for better governing principles, Woodward's engineers had gone to an emphasis on mathematical analysis as early as the 1940s. To aid them in this effort, company engineers had purchased, in 1949, the elements of a Philbrick analog computer and assembled them to meet their experimental requirements. The computer was extremely useful because it aided mathematical speculation in the control field in an exceedingly economical way.[38] Perhaps the greatest significance of the computer was that it showed that Woodward was staying ahead of its competition. Computers were unusual in the late 1940s, but Woodward had one and was making practical use of it.

Another practical method of promoting better governing principles was the company's decision to reinstitute Prime Mover Control Conferences (governor schools). These had been held every year from 1927 through 1931, then in the odd years from 1933 through 1939. World War II had caused their postponement until 1946. Thereafter, they were held annually, and delegates arrived from all over the world to mingle with

The "Open Office" concept is illustrated in this 1953 photo of Woodward's main office. At least six department managers are included in the area shown.

Mark Leum (left) and Dr. Rufus Oldenburger, Woodward mathematicians, used this Philbrick analog computer to simulate governor, engine, and load systems operation. This early 1950s computer used vacuum tubes.

bers had been with the company for decades, and the company was in good hands as Irl approached his sixtieth birthday in 1955. Old-timers like Walt Thorell, Roy Estell, Ward Ring, Art Nielsen, and Bill Taylor were stabilizing influences. New members, like Forrie Drake, Charlie Turner, Norm Noling, Urban Floor, and especially Bill Whitehead, seemed to be rising stars. Future company leaders like Cal Covert and Bruce Livingston were in their second decade of employment, with their influence yet to be felt.

The company's postwar success had exceeded even Irl Martin's expectations. Practically every year, Irl had advised caution in the members' handling of personal finances because surely the high profit sharing would not go on indefinitely. Irl felt the profit sharing system was a keystone to the democratic partnership philosophy of business association that symbolized Woodward's Corporate Partnership. The "relative value" rating system provided the basis for base pay in as democratic and accurate a way as could be found in a corporation. Each worker rated every other worker in his department on the basis of each person's relative value to the company. The ratings were averaged and listed numerically within the department. Each department did this and ultimately the ratings of all departments were reconciled by the plant Rating Committee. The final result listed the relative value of every member of the company, including management. On the basis of this list, each person was assigned a base wage rate or salary. Then surplus income was divided among the members. Longevity raised the profit sharing,

other delegates and Woodward members. The conferences continued to be a public relations bonanza for Woodward, since they provided a practical governing experience for the delegates and an inroad for potential sales.

Delegates who visited regularly in the postwar decade would see a remarkable continuity and stability in their Woodward Governor Company contacts. Many mem-

Governor School picnic on the front lawn of the plant, 1954.

"The Morning Mail Meeting"—a tradition at Woodward. The executives met ten minutes before regular starting time to read and "pass the mail." Later on, morning meetings were held in each plant to discuss current matters.

In this 1953 photo are (clockwise) Bill Taylor, Billy Bittle, Russ Martin, Bert Burridge, Bill Whitehead, Norm Noling, Tom Larsen, Art Nielsen, Joe Zircher, Roy Estell, Walt Thorell, and Irl Martin.

but the profit sharing percentage was the same for all members.[39]

During the decade, profit sharing (not including longevity) ranged from a low of about 22 percent in fiscal 1946 to a high of almost 44 percent in 1950. If a person or family lived on a member's wage or salary, as Irl counseled, the profit sharing was a welcome addition to discretionary income. The amounts were substantial enough to have a positive effect on Rockford's consumer economy, particularly in the purchase of big-ticket items like automobiles and home appliances.

At this point in the mid-1950s, a remarkable repetition of company history was in the process of occurring. Just as Irl was the son-in-law who was brought into the business and ultimately took it over, Irl had his eye on Billy Bittle, who had married Irl's daughter, Mary Barb, during World War II. Upon his discharge in 1945, Billy began working for Woodward, and his path was similar to the path Irl took in the 1920s. Billy started out as a stock chaser and then learned to operate a number of machines. Later, he was in Sales and Service. Like Irl, he was expected to learn the whole business through practical experience.

After Elmer Woodward's son, Harold (Doc), moved to California in 1951, Billy was named to replace him on the Board of Directors. By 1953, he was also manager of the Hydraulic Turbine Governor Department and, by 1955, he was assistant general manager. His rise would continue until 1960, when he became company president, an event that would mark a new and difficult chapter in the company's history.

Filming the first company movie, "The Woodward Way," in 1953. Woodward's Phil Kjellgren (far right) works with the film crew and actors.

Domestic and Foreign Expansion

1955-1965

Toward the end of Woodward's highly successful postwar decade, company leaders became convinced that their corporate philosophy ruled out further plant expansion at the Rockford headquarters site. Irl's ideal of a maximum of 500 company members under one roof had long since been surpassed, and he worried that the company's democratic partnership philosophy might thereby be eroded. Accessibility and trust were cornerstones of the Corporate Partnership. Common courtesy and personal respect, and the kind of informality that allowed members to call Irl by his first name, were the mortar that held the stones together. All of this required a relatively small plant.

Had Woodward remained a company that specialized only in hydro applications of its governors, smallness would have been assured. But expansion into diesel and aircraft governors in the 1930s, and the experiments on gas turbine engines during and after World War II, had propelled the company into a formidable, diverse organization whose prospects continued to improve. Thus, in the mid-1950s, the company decided to expand outside Rockford. The only questions were, when and where? Demand from domestic and overseas customers required additional facilities. This growth continued from 1955 through the following decade.

Another issue that influenced the company in this decade was company leadership. Irl was sixty years old in 1955, and a thirty-four-year member of the company. He was as dynamic as ever, even as his mobility became increasingly restricted by the deterioration of his hip joints. He eventually did have hip surgery, which was par-

Arizona Senator Barry Goldwater, shown here with Irl Martin, visited the plant in 1959. Senator Goldwater was the Republican candidate for President in 1964.

The Barber Shop and Recreation Area, about 1959.

tially successful, and was able to move around the plant in his "purple people eater"—a motorized purple golf cart. It enabled him to continue his informal face-to-face meetings with individual workers—an activity he felt was vital in promoting a continuing sense of camaraderie as views were exchanged.

Irl's greatest strength was his vision of the company and his ability to maintain it. But this also raised a problem: who would succeed him? And would his successor pursue the vision as zealously as Irl? Company executives, and Irl himself, would grapple with these questions for years to come, not always successfully. As they would come to learn, there was only one Irl Martin—and the times, and the people who came with the times, were changing. The company would weather these upper-management crises, but not without undergoing some intense soul-searching.

Irl Martin addressing members in a plant meeting, about 1959.

Despite these internal problems, this decade would see the company begin a period of accelerated growth and prosperity that would bring a high level of economic benefits to both worker and stockholder members. For the worker members, employment went from 769 at the end of 1954 to 1,430 in 1965; wages remained competitive, and the high semiannual profit sharing insured that the company continued to be seen by many as the best place in Rockford to work. New perquisites, such as the opening of the dental clinic in August 1957, added to the benefit package and enhanced the company's reputation. The stockholder members saw annual dividends climb from $1.60 per share in 1954 to $2.50 in 1958 and $3.50 in 1962. This annual yield was more than matched by the appreciation of the book value of the stock, which stood at $32.31 at the end of 1954. It increased every year, reaching $83.05 in 1965.[1] Those worker members who were also stockholders saw their investment in time, energy, and stock give them a double advantage. Anyone who bought stock during this period and held it for retirement would be amply rewarded.

As in the first decade after World War II, the years between 1955 and 1965 were prosperous ones for Rockford. Total population, which had dropped from 85,000 to 84,000 in the depression-ridden 1930s and was only 93,000 in 1950, grew rapidly to about 130,000 in 1960. By then,

Managers, officers, and consultants ready to serve Christmas dinner to members in 1957, a company tradition.

(From left) Roy Estell, Forrie Drake, Charlie Thomas, Bill Taylor, Art Logan, Dick Leighton, Honore Hubbard, Ernie Swarts, Bill Whitehead, and Norm Noling.

Rockford was Illinois' second largest city,[2] with a prosperity to match. Building permits set a new record in 1955 when they totaled over $19 million. More schools were built to accommodate the rapid growth in population. In 1955 alone, a record 4,552 babies were born. Businesses were also expanding. A $5 million American Chicle Company plant, a $2.5 million Elco Tool and Screw Corporation building, a $4 million plant for the Amerock Corporation, a $2 million addition to Sundstrand, and a new plant for the National Lock Company all were occupied in 1956 or 1957.[3] The building boom continued into the 1960s when Rockford avidly sought to attract a new Chrysler Corporation assembly plant. The attempt failed in Rockford, but the $50 million plant that employed 4,500 was built on the Northwest Tollway in nearby Belvidere.[4]

In the spring of 1957, Ben Schleicher was elected to the first of his three consecutive terms as mayor of Rockford (1957-1969). Irl Martin was especially pleased by this, because Ben was a Woodward Governor Company member. Ben's victories were an encouragement to others and showed that the company's emphasis on the civic duties of its members was having an influence on the community.[5]

Irl remained fully in control during this decade despite reaching the mandatory retirement age of sixty-five for officers in 1960. Although he had to retire as president, he retained his position as chairman of the board. He would continue as chairman until just before his death in 1976, and during that time four presidents would serve under him.

The first was Billy Bittle, Irl's son-in-law. His move upward in the company had been expedited by Irl himself, much as Irl's own rise had been overseen by Elmer. To many of the old-timers, this made a great deal of sense. Billy was

Irl Martin receives a warm welcome from Woodward members on his 65th birthday, December 12, 1960. This day marked Irl's retirement as president of the company.

Billy M. Bittle is inducted as fourth president of the Woodward Governor Company on December 23, 1960. Company officials and twenty-five year members are seated on stage.

Irl's interest in seeing Billy succeed was based on the relationship that had developed between the two men. In many uncanny ways, it resembled the working and personal relationship between Elmer and Irl. Both had married their boss's daughter, each of whom was her father's favorite. The Bittles also lived next door to Irl and Dorothy, which promoted close social contacts between the two families, including the four Bittle children. These close contacts carried over to the office, where history repeated itself. Irl and Elmer had sat at adjoining desks. Irl improved upon this by ordering a long desk which he shared with Billy. This caused a few tensions, as Billy learned that it was difficult to work next to his boss, who was also a living legend, without feeling at least a little self-conscious about the quality of his work.

family and Woodward was a three-generation family-run corporation that expressed family values as a major part of its Corporate Partnership. Another family member as president would provide stability and would further confirm the company's commitment to its explicit philosophy and traditions. Also like Irl, Billy was on a timetable that took him through most phases of the company's operations. He had begun with Woodward in 1945 in the shop, doing routine mechanical work. By the mid-1950s, he had gained experience in a wide variety of positions. He had been a company director since 1951. In 1953, he was made manager of the Hydraulic Turbine Governor Department and, in 1955 he was both assistant general manager and assistant secretary. By the late 1950s, he had been promoted from assistant to the president to general manager. He was named president in 1960.

Billy's appointment to the presidency was formalized on December 23, 1960, in a ceremony witnessed by the entire Rockford membership, visiting dignitaries, and Billy's own family. It concluded with Billy taking an oath of office to uphold the Woodward Way of doing business under the Corporate Partnership. Irl had requested the oath as a way of assuring himself that any successor would take the Corporate Partnership as seriously as he did.

The two men were remarkably different in ways that Elmer and Irl had not been. Billy was a good salesman, like Irl, but more of an extrovert. He could think on his feet so well that he sometimes anticipated, out loud, the response of someone he was speaking to. This could be disconcerting at times, and possibly interpreted as a sign that

Billy was not listening to their suggestions as attentively as Irl might. Billy was also cut more in the mold of the modern postwar manager, which required him to take a larger view of things than he could get by mingling so intimately with individual members of the company. Nor did he work in exactly the same way as Irl. He was not physically on the job as often as Irl wanted him to be. He had joined some local clubs and liked to travel, thinking, perhaps, that the company was well-run by a good staff of managers and supervisors, and it did not need his personal attention on every detail.

Irl and Dorothy Martin

These differences eventually reached a climax that led to Billy's resignation in 1965. But the clash of egos, expectations, and visions was as much a clash between two times as between two men: the hard years of the Depression and war set against the prosperous managerial years of the 1950s and 1960s, when America's greatness in the world was unchallenged, and its economic success a model. Both men were products of their times, and each was a worthy representative. But it became evident that what Irl expected of a president was only what he expected of himself, and that neither the times nor the man who was president promised another Irl Martin.

Billy's resignation affected different people in different ways. Many who had been disturbed by the growing tension between him and Irl were relieved that it was finally over. Others felt that Billy had been promoted too fast, into a position he had not been adequately prepared for. Few wished to interpret Irl's zeal for the Corporate Partnership as distrust of a subordinate. It was simply not Irl's way. But neither was anyone ready to accuse Billy of poor management. The general consensus was that the two men found themselves to be incompatible once they were brought so close together.

Billy and Mary Barb continued living next door to Irl and Dorothy for the next decade, although their relationship was never the same after 1965. It was proper, but the intimacy was gone.[6] Eventually, the Bittles moved to Colorado. Mary Barb had replaced Billy on the Board of Directors in 1966. She remained on the Board, insuring that the family continuity from Amos Woodward through four generations of management continued at the Woodward Governor Company.

While he was president, Billy was ably served by several subordinates. The three vice presidents, Walt Thorell, Forrie Drake, Bill Whitehead, and the secretary, Roy Estell, each had long careers with

Woodward. The treasurer, Dick Leighton, had joined the firm in 1956 and been named treasurer in 1959. Both Thorell and Leighton would later serve as president, but perhaps the most able of the five was Bill Whitehead.

Whitehead had joined Woodward in 1934 as an engineer. His own interests and capabilities were eclectic. He followed Irl's philosophy of learning the whole business, and the more he learned, the more valuable he became and the more responsibilities he took on. Colleagues described him enthusiastically as a "genius," a "tremendous ball of nerves and energy," a "pushing force," an "infectious, enthusiastic leader," a person with a "keen understanding of engineering principles," a guy who "could sell an idea and sell the product."[7]

Through 1958, he had served as chief engineer, sales manager, treasurer, and general manager. In 1959, he was appointed vice president in charge of sales and service. Since the company was operating plants in Europe by this time, he also took over the administration of all overseas sales and manufacturing. Whitehead seemed to thrive under pressure, and he pursued a hectic life with zest. No one can define charisma, but Whitehead was one of the fortunate few who had it. He was respected and well-liked by engineers and other company representatives around the world because of his contagious enthusiasm and his genuine interest in other people.[8] Coworkers saw him as an obvious choice for president if Irl chose to go beyond his family.

That did not happen, but he and Dick Leighton were personally friendly to Billy Bittle and supportive of his presidency. In fact, this trio, for the most part, ran management (under the watchful eye of Irl) between 1960 and 1965. The relationship was advantageous to Bittle because, in his absences from the company, he could be confident that Whitehead would run the day-to-day operations with a high level of competence. Unfortunately, Whitehead paced himself too hard. With "shocking suddenness" he died from a heart attack in Cairo, Egypt, on February 19, 1964, just short of his fifty-second birthday. He was just beginning a short vacation after visiting the company plants in Europe, before heading to Japan.[9]

The management trials which afflicted the company during these years did nothing, however, to slow down the company's expansion. Although business had expanded rapidly during the Korean War, Woodward had resisted expansion on the belief that the business stimulus would be temporary. This meant that, as in World War II, a portion of the business was subcontracted to other manufacturers. When it became apparent, however, that the increase in business looked secure for the foreseeable future, the company decided to expand the facilities in Rockford and to set up a new operation in another location. Irl's first choice was to find a building or building site in or near his hometown of Broken Bow, Nebraska. He thought he had found a probable location on an abandoned airport in southeastern Nebraska, but learned that all its buildings were committed for the storage of surplus corn.[10]

The search for an appropriate site continued through the autumn of 1954. After lengthy investigation, Walt Thorell was sent west to investigate two college towns—Laramie, Wyoming, and Colorado Springs, Colorado. Despite Irl's belief that higher education was often irrelevant compared to practical experience, others had convinced him that a college town would make a good site. Irl particularly wanted to locate in a small town with little or no existing industry. He was hesitant to move into metropolitan areas. They were simply too big and too impersonal, and he considered them potentially corrupting to the membership and not likely to have the strongest work ethics.

After visiting Laramie, Thorell paid a courtesy call on Irl's boyhood friend and baseball teammate, Dr. Fred Humphrey, in Fort Collins, Colorado. After Thorell explained he was on his way to Colorado Springs to consider it as a site for Woodward, Dr. Humphrey inquired, "Why not Fort Collins?" Colorado A & M College (now Colorado State University) was located there, and it was a small town with excellent prospects. Within hours, Humphrey arranged for the Chamber of Commerce to meet with Thorell. Humphrey quickly became the quarterback of a Chamber of Commerce game plan to attract the company. He phoned Irl, who balked, fearing that the taxes were too high, but who was interested on other counts. Within a very short time, Irl stopped considering any place other than Fort Collins and, by December 1954, the decision was made.[11]

FORT COLLINS COLORADOAN
Continuing Fort Collins Express-Courier
Woodward Company Invites Public to Inspect Plant
Flying Farmers of Two States to Convene Here
WELCOME and BEST WISHES
from a youngster of 55 years to a patriarch of 86
American
FURNITURE COMPANY
211 LINDEN STREET
We Are Proud . . .
that Woodward Governor Company chose Fort Collins for its latest plant expansion!
We Are Honored . . .
too, that the Woodward Governor Company selected our firm to furnish and outfit their new office with the very latest in desks, chairs, files, office machines, etc.
Remember . . .
CHECK WITH
Shedd & Lee
"Everything for The Office But The Stenerary"
Not "Whistle"—but MUSIC while you work
Woodward's Open House
8:30 to 11:30 a.m.–1:30 to 4:00 p.m.
Saturday, March 5th
Best Wishes to the Woodward Governor Co.
Zoric
COOPER MUSIC

Students Getting Valuable Experience for Life Work
Court News
Personals
We Welcome . . . WOODWARDS to Fort Collins
IVERSON DAIRY
We Hurried Over To Welcome Our New Neighbors To The East—The WOODWARD Governor Co.
Smith's Furniture Co.
We Welcome Woodward's to Fort Collins
ELECTRIC Shoe Shop, Inc.
Welcome—
WOODWARD GOVERNOR CO.
Our greetings, on the occasion of your OPEN HOUSE, Saturday, March 5th
May be anti-climactic, as
Fort Collins Really Welcomed You With
A BANG
When The Airport Bonds Carried Overwhelmingly
HI! PODNER!
Vogue . . . BOOTERY
"Style Without Extravagance"
You, Too, Will Love The Thrilling Atmosphere of Fort Collins
Guarantee Reserve Life Insurance Company
Home Office—323 South College — Fort Collins, Colo.

The March 4, 1955, edition of the Fort Collins Coloradoan *featured stories and advertisements welcoming the Woodward Governor Company on the eve of its Open House for the general public.*

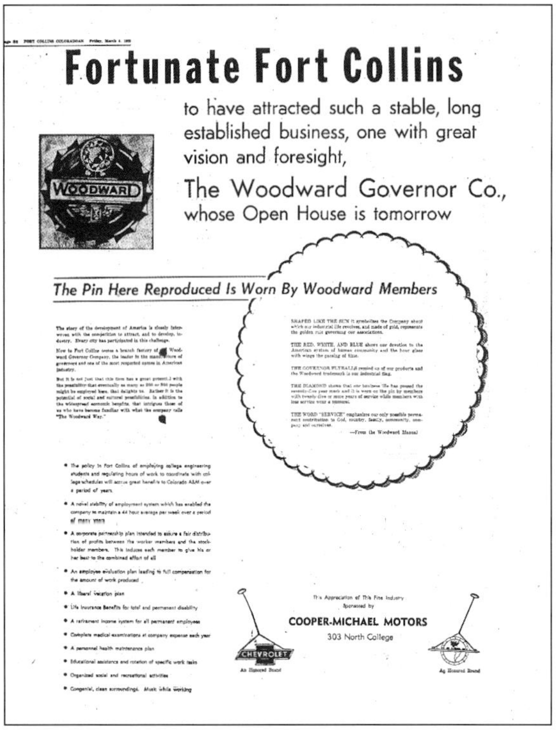

Fortunate Fort Collins

to have attracted such a stable, long established business, one with great vision and foresight,

The Woodward Governor Co.,
whose Open House is tomorrow

The Pin Here Reproduced Is Worn By Woodward Members

The story of the development of America is closely interwoven with the competition to attract, and to develop, industry. Every city has participated in this challenge.

Now to Fort Collins comes a branch factory of the Woodward Governor Company, the leader in the manufacture of governors and one of the most respected names in American industry.

But it is not just that this firm has a great potential with the possibility that eventually as many as 200 or 300 people might be employed here, that delights us. Rather it is the potential of social and cultural possibilities, in addition to the widespread economic benefits, that intrigues those of us who have become familiar with what the company calls "The Woodward Way."

SHAPED LIKE THE SUN it symbolizes the Company about which our industrial life revolves, and made of gold, represents the golden rule governing our associations.

THE RED, WHITE, AND BLUE shows our devotion to the American system of human community and the hour glass with wings the passing of time.

THE GOVERNOR FLYBALLS remind us of our products and the Woodward trademark is our industrial flag.

THE DIAMOND shows that our business life has passed the seventy-five year mark and it is worn on the pin by members with twenty-five or more years of service while members with less service wear a [illegible].

THE WORD "SERVICE" emphasizes our only possible permanent contribution to God, country, family, community, company and ourselves.

—From the Woodward Manual

- The policy in Fort Collins of employing college engineering students and regulating hours of work to coordinate with college schedules will accrue great benefits to Colorado A&M over a period of years.
- A novel stability of employment system which has enabled the company to maintain a 44 hour average per week over a period of many years.
- A corporate partnership plan intended to assure a fair distribution of profits between the worker members and the stockholder members. This induces each member to give his or her best to the combined effort of all.
- An employee evaluation plan leading to full compensation for the amount of work produced.
- A liberal vacation plan
- Life Insurance Benefits for total and permanent disability
- A retirement income system for all permanent employees
- Complete medical examinations at company expense each year
- A personnel health maintenance plan
- Educational assistance and rotation of specific work tasks
- Organized social and recreational activities
- Congenial, clean surroundings. Music while working

This Appreciation of This Fine Industry Sponsored by
COOPER-MICHAEL MOTORS
303 North College

Company Founded in 1870

Welcome
The Woodward Governor Co.
Greetings from the folks at Bowen's
Greeting Cards A Specialty
Your store of friendly service
Bowen's Book Store

WOODWARD—an old Institution with Modern Ideas is welcomed by FORT COLLINS an old town that is going progressive.
Sincerest congratulations and best wishes—from
Klinker-Keese Service

We Feel Fortunate
—also good—to have the
Woodward Governor Co.
in our community
Our Warm Personal Greetings and Best Wishes are extended to you as you hold
OPEN HOUSE
at 337 Jefferson
—from 8:30 to 11:30 a.m. and from 1 to 4 p.m. tomorrow.
Sincerely,
ED. BOWLING
STARK'S

We Hail...
The Woodward Governor Co.
Building for the Future
in our fine city
Mawson Lumber and Hardware Co.
307 S. Mountain — Phone 774

2 Withdraw Backing for Water Board Bill

Rep. Markley Would Hike Gas Tax or License Fees

Hits Tender Spot

We are fortunate to have
The Woodward Governor Co.
bring 85 years of
engineering EXPERIENCE
to Fort Collins.
The D. C. Royer Agency
offers
42 years of Experience
in the Insurance Business
The D. C. Royer Agency

PAINT
used in refurbishing the building of the Woodward Governor Co. at 337 Jefferson was furnished by us.
Sherwin-Williams . . . Paint Store
221 North College

Adding Strength to Colorado A&M
always one of Fort Collins' greatest assets

Woodward Governor Co.'s Policy of Employing
Colo. A&M Engineering Students

We Join
With the Rest of Fort Collins
In Extending
The Heartiest of Welcomes
To Woodward Governor Co.
Robinson-Warfield Printing Co.
Phone 314 — 142 W. Oak St.

Art Nielsen was named plant manager, and seventeen machine operators were sent with him in early January 1955, to set up operations in a rented building that previously housed the Valentine Motor Company. The machine operators set up screw machines and turret lathes and provided instruction for the fifty-six Colorado A & M students who were hired to work at the plant. Initially, the plan was to have them work a regular eight-hour shift two or three days a week while they continued taking classes. Most of these young men were in school to become mechanical engineers, and their part-time work enabled the company to evaluate their potential for full-time jobs after graduation. In the meantime, they performed blanking operations on parts made from steel bar stock for use at the Rockford plant.[12]

Within three years, Woodward occupied three buildings (two of which they owned) in Fort Collins, and the small auxiliary operation was growing by leaps and bounds. Over 250 people were employed by 1958, about half the optimum personnel size for a facility. Students played an important role in the plant as expansion continued and many of the original work-study students became company members. By late 1956, both a day and a night shift were working forty-five hours each, with students making up a substantial part of the night crew. From a machine shop that began by performing the initial operations on steel parts, the Fort Collins branch expanded into making complete governors by 1957. All parts were manufactured, assembled and tested there.[13]

Fort Collins was supposed to show a profit in five years, but this was achieved in only three-and-one-half years. It was no easy task for many of these "cowboys and cowgirls" to make the transition from ranches and farms, where they could wear jeans and colorful shirts, to the more regimented Woodward Way. People were used to wearing jeans even to church, and now here were

The former location of the Valentine Motor Co. automobile dealership, this vacated building (right) at 337 Jefferson Street was remodeled by Woodward Governor Company to house the company's first Fort Collins operations in early 1955. By 1958, Woodward had expanded into the adjacent Schneidmiller building (left) and the King building (far left) across the street. In 1960, a cafeteria connecting the two adjacent buildings was constructed.

some "Easterners" telling them to wear coats and ties to work. Some left, but those who stayed adapted quickly. Management soon learned that the typical "Westerner" was used to hard work and could hold his own very well in production rates with the people in Rockford.[14]

By 1960, Woodward's operation in Fort Collins consisted of two adjacent buildings and a newly-built cafeteria. By the early 1960s, further expansion was necessary and plans were made for the construction of a new plant on a site away from the city center. Forty acres of land were purchased in 1962. By early 1965, a "pilot building" was completed to begin operations in machinery, painting, plating, coating, storage, assembly, test, and shipping. A grounds maintenance building was completed later in 1965, and plans went ahead for the main building. When it was completed in September 1966, all manufacturing operations moved from the downtown location. The old buildings were later sold.[15]

By 1965, the size of the membership in the Rockford plant was almost 1,000—well above the company's target of 500. Consequently, volunteers were solicited to move to Colorado through a low-key recruiting program that touted the advantages of living in Fort Collins. The company promised aid in arranging transportation and selecting living accommodations.[16] As a result, a number of Rockford members relocated to Colorado.

During this decade, the company was also moving to expand to Europe, Great Britain, and Asia. Bill Whitehead and other company members tried to convince Irl that Woodward could retain and expand its European business only by establishing a branch on the Continent. In early 1955, Roy Estell, Woodward secretary, was sent to the Netherlands to make the first personal customer contacts. He was encouraged enough that Irl and company treasurer Charlie Turner traveled to the Netherlands in mid-September to meet with customers and Dutch government officials. Irl was interested in the immediate need for sales, service, and repair work, and the possibility of later adding manufacturing. On the other hand, he wondered whether Woodward might be getting too greedy, and he was generally concerned about the difficulties of setting up a company in a country where the language was not English and the laws and customs were unfamiliar.[17]

The trip was a success from Irl and Charlie's point of view. They traveled throughout Europe and met with government officials in Great Britain and the Netherlands. Through earlier investigation, they had determined that the best possible location would be in or near Amsterdam, preferably at Schiphol, Amsterdam's international airport. With Dutch cooperation and encouragement, a room in Hangar Five was rented in December, and sales and service work began soon after.[18] Company expansion into England was put on hold until 1958, when a separate operation became necessary because of the trading patterns of the Common Market that initially excluded Great Britain and Scandinavia.

In April 1956, more space was rented in Hangar Five and assembly operations began. Additional space was added in 1959, 1962, and 1966 for offices, a machine shop, storage, receiving, inspection and stockrooms. Land was purchased in nearby Hoofddorp in May 1964 for a future plant. Construction on the new building began in February 1966, and it was completed on March 14, 1967.[19] By 1958, the Dutch plant's potential convinced Irl to make plans to expand to the United Kingdom and to order a survey for a Schiphol-type operation in Japan.[20] The operation in England started later in 1958, and the one in Japan began in 1960.

Assembly and Test area in Woodward's first Slough plant, about 1960. Members are (from left) Harry Simpson, Ray Johnson, Mike Nagle, and Bob Lantz (from Rockford).

Jack Shlikas had replaced Roy Estell as manager of the Netherlands plant in January 1957, and Woody Carson arrived in January 1958 to assist him for a year before becoming manager for the 1959-1962 period. By Rockford standards, this first overseas operation had started out quite primitively because nobody knew much about overseas management or overseas needs. But the move was a good idea, and people struggled together to solve problems and to learn from mistakes. Governors built by European competitors did not match Woodward's in the 1950s, so the company had a distinct advantage if it could build customer good will. The fact that parts and some governors bore the label "Product of Holland" aided European sales.[21]

By mid-1958, fourteen Dutch citizens and four Americans were running the operation at Schiphol Airport, producing parts, assembling and testing governors, and performing sales and engineering application functions. Because Woodward's Corporate Partnership philosophy was so different from the business standards people in the Netherlands, England, and Japan were accustomed to, Irl and other company officials made regular visits overseas to explain it and to convince new members that it was to everyone's advantage to adopt these proven ideals.[22]

In England, the company opened sales and service offices, first in Bolton (near Manchester) and then in London. By late 1959, sales, service, assembly, and tests were carried out in a new, leased plant on an industrial estate in Slough, about twenty miles west of London and convenient to Heathrow Airport. An adjoining building, which was leased in 1962, doubled the original work area. Through leases in 1967, the company doubled its size again by 1971.[23]

In Japan, an office opened in Tokyo in 1960. After growing out of its initial location and a second location, operations moved to a 5,340 square foot second floor location in 1964. In 1967, the first floor was leased, nearly doubling the space. The company remained there for a decade.[24]

Thus, by the end of the 1955-1965 decade, Woodward was well established in its four branch plants in Fort Collins, the Netherlands, England, and Japan. Irl was pleased with the results and proud that the local domestic company of 1955 had turned into such a strong national and international company without compromising its standards or beliefs. The Netherlands plant was in the black by 1962, and England and Japan were expected to follow suit by the mid-1960s. Irl, who had been a reluctant convert to expansion in 1955, was by now an enthusiastic supporter, saying, "I find the kids at Slough and Schiphol are just the same as we are here—honest, hardworking and loyal."[25]

As painful and disruptive as expansion could sometimes be, it did strengthen the company, and the product history during this decade also was diverse enough to allow for expansion without threat of serious financial risk-taking. The product leader during this decade was Woodward's 1307 fuel control for General Electric's gas turbine engines. By 1955, work on this control had been going on for years and, in late 1956, Urban Floor, the chief development engineer, noted that:

Japan branch members pose in front of Woodward's first location in Tokyo, the Fujiya Building. (From left) front row, Mimi Matsumoto, Tsuruzo Asami, Eiko Tada; middle row, Takashi Abe, Walt Thorell, Takeo Yuasa, Motoichi Kurashige; back row, Hirokazu Komatsu, Tat Hongo. The photo was taken about 1962.

> *As in the past several years, our major effort has been concentrated on the development of fuel controls for aircraft propulsion turbines. This policy is now beginning to yield returns, as two large jet engine manufacturers have placed orders for initial production quantities...*[26]

He went on to mention two controls, the 838 control for Westinghouse, which was modestly successful in sales, and the superior 1307 control for General Electric. The 1307 enabled Woodward to solidify and improve its ties with General Electric and, more importantly, to

Faye Sweeny (left) and Paul Challberg are shown with the first X1307 jet engine control that was shipped to General Electric on January 25, 1957. Woodward continued building these controls for over thirty years.

get its controls installed on a wide variety of military and commercial jet aircraft.

Work on these controls was a cooperative venture involving a number of engineers, although Urban Floor gives primary credit to Harland Plummer for designing the 1307.[27] The first one was shipped on January 25, 1957. Initially, they were expensive to make because of their complexity. Assemblers took as long as 100 hours to put one together, and it might take another 40 to 100 hours to get it through test. In an effort to simplify the assembly process, Ward Ring asked the foreman of the Diesel Department, Rex Hunt, to take over the 1307 project. Hunt developed a six-station sub-assembly network of specialized assembly tasks, and he built into the assembly process progressive inspection under a checkoff procedure. (The six stations were eventually changed to four.) With this new process, assembly time for the 1307 dropped to about twelve to fifteen hours. Because the inspection system was also done in stages, there were fewer delays, and testing dropped to the six-to eight-hour range. As a result, profits skyrocketed—too much and too fast, in fact.[28] Given the company philosophy of striving only for a fair profit, Woodward returned a large chunk of the excess profit to General Electric which, surprisingly, asked Woodward never to do this again.[29] Getting rebates of this sort had put G.E. into an awkward position when subcontractors' costs were lower than those the contract called for.

By 1959, the 1307 was being applied to a variety of military and commercial projects. It was on General Electric engines on the B-58 Hustler, the A3J-1 Vigilante, the F-104 Starfighter, and the Convair 880 passenger plane. In each case, the fuel control was considered reliable and capable of high performance.[30] Later that year, the company proudly promoted the Air Force's upper atmosphere research plane, the X15, which was equipped with Woodward's variable speed hydraulic governor. This installation on the rocket engine of the X15 meant that Woodward's experience in speed control of prime movers extended from under the sea to the edge of space.[31]

By 1961, the 1307 was on the French-built Caravelle VII passenger plane and the McDonnell Douglas F4H Phantom II, a Navy fighter plane. The Phantom, Hustler, Vigilante, and Starfighter held a total of eighteen world aviation records for speed, altitude and time-to-climb marks. This meant that planes that had the 1307 on them held one-third of all world class records. By this time, too, the Convair 880 was considered the world's fastest jet liner, and a number of companies, including TWA, Delta, and Japan Air Lines had adopted it.[32]

F4H Phantoms continued to break many records, and each of the military planes with Woodward components began to play a larger part in the defense of the United States, especially as the country became more deeply involved in Vietnam in the early to mid-1960s. F-104 Starfighters were also purchased by America's allies in Europe, most notably the West Germans. Consequently, sales for these controls boomed in the 1960s as the United States and her allies invested heavily in military hardware.

One other practical, and unusual, adaptation of the 1307 was accomplished through the voluntary labor of a number of Woodward members. They had rebuilt, on their own time but with management's approval, a 1307 that professional race car driver Art Arfons had brought to them needing a major overhaul. Arfons had earlier purchased a used J-79 jet engine with a 1307 control and was adapting it to fit his race car, "The Green Monster." With the members' free help, he steered the "Green Monster" to an "amazing 536.7 mph average" speed on the Bonneville Salt Flats in Utah.[33]

By 1965, the 1307 had been the flagship of Woodward controls for more than five years. Each year in the first half of the 1960s, the Large Gas Turbine Controls Department touted the success of the 1307. In practically every year, a new sales record for this control was set. Even by the mid-1960s, when a newer control was described as the "planned successor" of the 1307, sales continued at a high level so satisfactorily that the company could say:

It still remains a major contributor to the Woodward Governor Company sales picture. Applications of 1307 controls for new versions of the General Electric J-79 jet engines continue to be our most promising prospect for near future business.[34]

Jet fuel controls were preeminent by the 1960s, but other departments within the company also had breakthroughs during this period. For example, in 1957, the Development Engineering Department logged a new project almost every working day. By December, 120 projects were on the books. By this time, the load sensing governor (LSG-1) had been released to the Diesel Production Department. This was Woodward's pioneer product using electric controls. This governor sensed load changes electrically and it sensed speed changes mechanically.[35]

In 1958, sales of diesel governors dropped, in part because of strikes at some of the engine plants, but also because the diesel locomotive field had begun to play itself out. Nevertheless, Arnie Swenson, manager of the Diesel and Steam Governor Department, was optimistic about 1959. The LSG was expected to go into full production, and its purpose, to provide a more precise regulation of electric power, seemed pertinent to new military uses, such as missile tracking radar. Chief Production Engineer Tom Larsen agreed, noting that the LSG was "being applied to an ever-increasing list of diesel engines for various applications and [was] being enthusiastically received." How enthusiastically was revealed by Jack Shlikas, now manager of the Diesel Department, who noted that LSG governor sales "jumped markedly" in 1959. It was especially useful in controlling remote or isolated stations like the Distant Early Warning (DEW) line across Alaska, Canada, and Greenland.[36]

Between 1960 and 1962, the LSG was in the process of being superseded by a new development—the Electric Governor (EG). This governor was under test in 1961 and in production by 1962. Urban Floor, the chief engineer, was confident that the EG would soon be applied to all kinds of prime movers where precise control was required. By 1963, the EG was being used successfully all over the world, and it was easily superseding the LSG by 1964. Sales of both continued to be important in 1965.[37]

In the hydraulic turbine field, electric governors began to make a major impact during this decade. Automatic equipment was increasingly desired, and the electric-hydraulic actuator permitted that. It was under test in 1957 and in production during 1958. In March 1960, the first electric-hydraulic actuators were scheduled for shipment to Ontario Hydro's Red Rock plant, which went into service later in 1960.[38]

During 1960, Woodward had two electric-hydraulic actuators on the production floor and sixteen on order. They were an instant success and a major improvement over the old actuators. By 1962, water wheel governor sales increased 66 percent over 1961—with electric-hydraulic actuators responsible for a significant part of the increase. Sales continued strong through 1965, and "interest and activity in electric-hydraulic" actuators increased rapidly.[39]

During this decade, company governors had a wide variety of unusual applications. A standard PG-PL governor was adapted for use on gas turbine electric locomotives that were the most powerful in existence. LSG-10 governors on diesel engines supplied power for Atlas missile launching sites; UG32 governors were on marine installations, including the "Zwarte Zee," the world's fastest tugboat; and Woodward SG governors and 1173 fuel controls were on an unusual thirty-five-ton Swedish amphibious tank. LSG governors were on Nordberg engines in Titan I missile complexes and at the North American Air Defense Headquarters deep underground in Cheyenne Mountain near Colorado Springs, Colorado.[40]

Just as satisfactory as product development were improvements in sales, profits, pay, and profit sharing. At the end of 1954, Woodward had employed 769 worker members, all located in Rockford but, by the end of 1965, its 1,430 worker members were dispersed through five locations. Rockford employed 1,040 (the highest since World War II), and Fort Collins 242. The overseas plants had 68 in the Netherlands, 43 in England, and 37 in Japan. Each of the overseas plants was expanding slowly, while Fort Collins had fewer worker members than it had in 1960.[41] However, each plant would expand more rapidly beginning in 1966, in large part because of the demands of the war in Vietnam on Woodward products.

Sales were in the twenty- to twenty-eight million dollar range from 1957 through 1965. This allowed for solid profits each year that ranged from $5 per share in 1960 to slightly over $10.50 per share in 1963. Profit sharing hit a low of 25.1 percent in 1960 and a high of 52 percent in 1962, the first year the company earned over $10 per share.[42] The average base pay remained slightly below the average for the Rockford area. This would help prevent layoffs in weak years, while in strong years the high profit sharing would more than make up for the deficiency in base pay. This conservative approach was part of the philosophy of the Corporate Partnership. Not surprisingly, considering how well-managed the company was, layoffs were never a problem in this or any previous decade, and profit sharing always made Woodward members the envy of other workers.

The Woodward Way continued to evolve during this decade. Education of new worker members remained a prime concern, particularly as the company expanded. New and potential employees in Rockford learned about company traditions and the Corporate Partnership. It was much more difficult to establish a similar tradition in Fort Collins, and doubly difficult to explain these highly unusual principles to citizens of the Netherlands, England, and Japan. Language barriers were problems in the Netherlands and Japan, and a long tradition of worker-management conflict in England was difficult to overcome.

Except for the addition of a company dentist in 1957 and some pension improvements that benefited the worker members, the only significant addition to the Corporate Partnership was the inclusion of a Recruit Training Program in 1960. Woodward had already been hiring high school and college students for summer

These recruits were in the first two classes, in 1960 and 1961. The Yard Maintenance Supervisor is Cal Covert (standing, second from left).

Irl's 70th Birthday.

yard work for a number of years. But, in 1960, the company set formal procedures for hiring students and established specific duties for their summer jobs. The plan was to find people with aptitudes and attitudes compatible with the company's purposes. Theoretically, this new talent could then be molded into valuable members.[43]

Boys fourteen years old who had completed the eighth or ninth year of school were encouraged to apply. Those who applied were screened through mechanical and technical aptitude tests, personal interviews, references, and comprehensive health examinations. In time, the program became quite elaborate. It envisioned dedicated young men who would graduate from high school with up to four summers of work experience at Woodward. Then, they would be offered full-time employment in the hope that they would commit themselves to the company for the foreseeable future. The recruits, by the time they were eighteen, would thus have received valuable training through yard work for at least two summers and then practical shop training for the next two summers. Those who chose to go on to college would continue to work summers in positions that were as relevant as possible to their college programs. On graduating they would have earned two years of Woodward seniority. The recruits also attended classes and lectures on a variety of topics, from care of equipment to the history of the company.[44] The idea fit in well with Irl's often-stated view that a practical education was just as important as book-learning or higher education.

Cal Covert, who was then in charge of the grounds crew, directed this program in the early 1960s. Under his guidance, it taught practical work habits, methods of getting along with others, and self-discipline. It was not just on-the-job training. The program received a Freedoms Foundation Award in 1964 for its accomplishments in helping young people achieve a better understanding of the American way of life.[45]

Company leaders believed that a retention rate of 20 to 30 percent would justify keeping the program, but they also believed the training was worthwhile even for those young people who did not stay with Woodward. Irl hoped that a recruit program could be established at each plant and that regular exchanges of recruits among plants might be possible. In fact, the latter hope took root in 1965, when four recruits from Rockford spent the summer of 1965 at the company's branch in the Netherlands.[46]

Irl passed his seventieth birthday in December 1965, proud of the Recruit Program and firmly in control of the company. Because of policy differences with Irl, Billy Bittle resigned the presidency that month and Walt Thorell was selected as an interim president for a year. Meanwhile, Forrest Drake, then vice president for Engineering and Research, and a company member since 1938, resigned his vice presidency to move to California, where he intended to continue doing full-time research for the company.[47] The Board of Directors, which Irl chaired, voted to waive the company's rule for mandatory retirement at age seventy, enabling him to remain as chairman. Irl then insisted that he be declared ineligible to receive any profit sharing (although his salary did continue), and the Board agreed.

The company thus began what would be the last decade of Irl's life in 1966. This next decade would see a number of personnel changes within the company and continued turnover in the office of president. It would also see the company facilitate minority hiring to match the stated principles of its Articles of Association, which promised membership "to all who have the necessary qualifications regardless of race, creed or color."[48] All in all, the 1955-1965 decade would serve as a springboard for continued success in the next decade.

The End of an Era But Not of an Idea

1966-1976

Irl Martin was not going to allow old age and its infirmities to slow him down. In 1965 he turned seventy years old. In the following decade, he would undergo hip replacement surgery, a rare treatment in those days. He would also develop heart disease and cancer. For most of that decade, he lived in semi-retirement in Yuma and then Phoenix, Arizona. But he remained as articulate and as firmly principled as ever. His enthusiasm for the Woodward Way and the Corporate Partnership never waned. He firmly believed, and with good reason, that he could guide, protect, and codify his ideas so that they would remain with the company after he was gone. But he wasn't about to go yet. He still ran the company. Although three different presidents would serve under him, each with differing degrees of freedom, his strong constitution would see to it that he remained as chairman literally until he was too ill to carry on.

This decade was also a difficult period for the United States. Changes in economic structures and race relations, the war in Vietnam, and the government crisis over Watergate remapped the surface of American society.

The war was the most obvious catalyst for change. It had begun in the late 1950s as a small police action involving only a few American advisors to the army of South Vietnam but, by the mid-1960s, it had become a war larger and more costly than Korea. By 1968, half a million Americans were serving in Vietnam. As the conflict dragged on, American society became fragmented over the question of involvement. Some people wanted complete withdrawal; others pressed for even heavier participation so long as it would lead to a military victory. The present stalemate was unacceptable to both sides. Then the Tet offensive of 1968, whatever its inconsequential military results for the North Vietnamese, underlined the hopelessness of the war. It became

Irl C. Martin

Billy Bittle

obvious that the war could not be won without a greater investment than the country was willing to give.

President Johnson also faced difficulties at home. He had inherited President Kennedy's incomplete plans for economic and social reform. They had grown out of the awareness that America was opening itself to those groups that had previously been left out of the mainstream of social and economic opportunity. Thus, while Johnson was prosecuting the war abroad, he was also prosecuting his "War on Poverty" at home. He envisioned a "Great Society" as an extension of Franklin Roosevelt's New Deal. The Great Society would address social and economic inequities that had not been cured by the prosperity of the 1950s. It began in 1964, with the passage of the Civil Rights Act, and continued through 1965 with the Voting Rights Act and further refinements to the Civil Rights Act.

Reform was slow, however. Social unrest over Vietnam, poverty, and racial discrimination grew into riots. It was clear, too, that Johnson's attempt to manage both the war and social change was not going to work out as he had hoped. Tet had shaken his confidence that the conflict could be ended on terms favorable to South Vietnam, and the cost of managing both the war and social change was more than the country's economy could bear. By the 1970s, it would produce an unusual combination of inflation and recession. Before that, however, it would doom Johnson's bid for re-election in 1968.

He was replaced by Richard Nixon, who called his plan to end the war "Vietnamization," or turning the conflict over to the South Vietnamese military while American troops were slowly withdrawn. Meanwhile, Nixon kept the pressure on the North by continued bombings and, at one point, mining the harbor of Haiphong. Finally, in January 1973, a peace treaty was signed in Paris and the war was, for America, over.[1]

At home, Nixon sought to bring an end to economic troubles through a combination of price controls and credit restrictions, with only moderate success. He also reopened talks with China and pursued detente with the Soviet Union through arms treaties and cultural exchanges. He was easily reelected in 1972, at the height of his popularity. But it all ended with the Watergate scandal, a crisis as damaging to the presidency as to Nixon himself. It forced him into resignation, in the face of a threat of impeachment.

After Billy Bittle resigned in December 1965, Irl turned to Walt Thorell to serve as interim president. Walt, who had joined the company in September 1920 (just before Irl), was sixty-four at the time. The assumption was that he would be a one-year president, stepping down when he reached sixty-five to make room for a new "permanent" president. The honor, nevertheless, of serving as president even for one year was a fitting capstone for a career that had spanned more than forty-five years. A native of Rockford, he had begun as a draftsman, but also had worked in the shop as a machinist and assembler. He spent a couple of years at North Park College in Chicago, and also picked up some engineering experience working part time with Harza Engineering Company. When he returned to Woodward, it was to work in a new capacity as a service and product engineer. In 1939 he was made manager of sales and service and also became an officer of the company as assistant secretary. The next year he was made secretary and, in 1942, he was appointed vice president. Between 1941 and 1959, he served in a number of other management positions, including production manager, purchasing agent, and manager of physical facilities.

In 1960 he was sent to Japan, where he spent the next twenty-seven months helping to set up the new assembly plant and sales and service network. When he returned from this venture, he was made executive vice president, which included responsibility to oversee the new data processing section. After a career like this, which had witnessed Woodward's rise to international prominence from humble beginnings, Walt was understandably nonplussed by the title "interim" president. From his perspective, he was making fast progress toward becoming a good president when he reached sixty-five. He wished to continue on, but Irl would not agree to this.[2]

Dick Leighton (left) accepts presidency of Woodward from Irl Martin.

Walt's year as interim president allowed Irl to prepare the next man who would be president, Dick Leighton. Leighton spent the year under Walt's wings going through an informal management succession program. Prior to this, Leighton had spent all of his ten years at Woodward in Accounting. His rise to the presidency was as spectacular as it was

Walt Thorell

Twenty-one members were inducted into the Twenty-five Year Club in March 1966. The event was held on the Shop floor because, due to lack of space, engineering functions had been temporarily moved to the Auditorium.

In the photo are (from left) Chaplain Haves White, Arnie Swenson, Gordon Beckman, Clarence Munson, Harold Drevs, Norman Resh, Dale Wetzel, Charles Shepherd, George Stassi, Rudy Swanson, Barrett Smith, Lit Schulein (partially hidden behind Barrett), Harley Bohl, Ken Douglas, Paul Abramson, Don Wick, Dave Kempe, John Lindvall, Herb Johnson, Bill Gaede, John Brewer, Bill Ragnar, and Walt Thorell.

short. He had joined the company in 1956 as an accountant working under company treasurer Charlie Turner. He served for two years as assistant treasurer under Bill Whitehead following the tragic 1957 death by leukemia of forty-six-year-old Charlie Turner.[3] In 1959 he was appointed company treasurer. During Billy Bittle's presidency, Leighton demonstrated his even-handed fairness and integrity by keeping loyal ties to both Billy and Irl. In March 1966, at age thirty-eight, Dick was promoted to vice president and general manager.[4] He was clearly the heir-apparent president, in need of only one more year of tutelage, until Walt retired at the end of 1966.

Irl believed that he had found the right man in Dick Leighton. He was hard-working and fair, and also believed in the Woodward Way and Corporate Partnership.

Although Irl remained active in company affairs, ostensibly to help Leighton grow into the job as president, he gave Leighton more freedom than either Billy or Walt had enjoyed. The long double desk, once shared by Irl and Billy, was removed, and Irl himself took up station in a separate office. During Leighton's first few years as president, he and Irl formed a close, trusting working relationship.

Leighton's increased authority and greater operating freedom, and his tendency to entertain ideas without seeking Irl's consultation, began to erode their relationship. Much younger than Irl and eager to expand the business, Leighton practiced a management style that sought to make changes faster than Irl believed was necessary and prudent. Irl's management style was increasingly cautious as he feared rapid changes would threaten the Corporate Partnership philosophy. Leighton was interested in branching out into product lines that were not strictly limited to governor and prime mover technology. While they might promise high profits, they were also riskier than Irl believed the company's business philosophy allowed.

Finally, the rift between Irl and Leighton widened until all hope of reconciliation was lost. Leighton resigned in October 1972.

Succeeding Dick Leighton as president was Bruce Livingston, a company member since 1941 and the quintessential Woodward man—someone who had worked himself up from laborer to president on ability and hard work. Livingston had held a number of positions in inventory and production control, and had served as manager of domestic sales, assistant secretary, Rockford plant manager, and vice president, before being chosen president. His tenure as president would encompass a full decade—including the first six years after Irl Martin's death in 1976.

The decade between 1966 and 1976 saw strong markets in prime mover controls, although it also had to face painful economic choices during the aerospace recession of the early 1970s. The strong financial results of this period built upon the inventions and financial successes of the previous decade. While net billings for products and services more than doubled from $32.7 million in fiscal year 1966 to $75.9 million in fiscal year 1976, the increase was pronounced from 1972 through 1975. In 1968, the year of America's heaviest involvement in Vietnam, billings reached $42.3 million. They declined in 1969 and 1970 before nudging minutely upward in 1971.

Woodward was not the only company in the aerospace business that faced tough times. In 1971 and 1972, the entire industry was hit with a massive slowdown in military and commercial aviation. The war in Vietnam was winding down, and a reduced defense budget meant fewer

Bruce Livingston, then general manager of the Aircraft Controls Division, talks with a TWA captain in front of one of the General Electric CJ-805 jet engines which power the Convair 880 airliner. This plane was on display at Rockford Airport, with the sign referring to the Woodward 1307 fuel control on the engine.

orders for military aircraft components. Commercial aviation had reached a plateau when production of the first generation of passenger jets, the Boeing 707 and 727, and the McDonnell Douglas DC-9, began to slow down. Although newer and larger jets were being built—the Boeing 747, the Lockheed L-1011, and the McDonnell Douglas DC-10—not enough orders had been placed by the airlines to make up for the decline. Thus, the first major aerospace recession hit, and with it came layoffs throughout the industry. While Woodward remained profitable throughout this period, the Corporate Partnership philosophy was strained when the Board of Directors decided that some Rockford members would have to be laid off in 1970. This decision was not arrived at easily; it would be the first time since the

Woodward's Walt Thorell demonstrates the marine synchrophaser to U.S. Ambassador W. R. Tyler (left) and Prince Bernhard of the Netherlands, at the Hoofddorp plant dedication, in 1967.

Corporate Partnership went into effect in 1946 that members were laid off. In the normal course of events, natural attrition would reduce membership if reductions were needed, but the depressed market conditions in 1970 required more drastic action, even though the decision was unpleasant and unpopular. In keeping with corporate philosophy, value to the company, not seniority, was the criterion used to decide who would be laid off.[5] Thereafter, employment wouldn't pass its 1968 high of 1945 worker members until 1974—a year when billings and unfilled orders were multiplying.[6]

During this decade, the company still had only 190,000 shares of its stock in circulation despite some stockholder pressure to split the stock.[7] Even the 50,000 shares that had been held in reserve since reincorporation in 1940 remained unissued. Because the number of shares in circulation remained relatively small compared to billings, profit, and working capital, the shareholders' book value per share went from $91.29 in 1966 to $218.62 in 1976. Regular and extra dividends also augmented the stockholder members' stake in the company. Dividends went from $2 a share in 1966 to $6.75 in 1976.[8]

The profit sharing for worker members varied considerably during this period. It started out at 23.8 percent in 1966 but declined each year until it bottomed out in 1970 at 11.0 percent. There was no spring profit sharing that year because the company was not making a profit, although it would show a small profit by the end of the fiscal year. The next year the profit sharing was only 12.2 percent. By Woodward standards, these were inadequate.

Thus, the most sensitive barometer of the Corporate Partnership's success, compensation, was undergoing a severe trial. In good times, the salary would be supplemented by generous profit sharing. In bad times, hours would be pared, subcontracting reduced, and profit sharing would be cut or eliminated before anyone was laid off. But not even these measures prevented layoffs in 1970. Fortunately, after these two very mediocre years, profits increased and profit sharing improved extensively to 22 percent in 1972, to over 40 percent in 1974, and to over 50 percent in both 1975 and 1976. In these latter years, substantial deferred profit sharing was also paid.[9]

The economic difficulties caused by the aerospace recession are reflected in the company's sales figures for the decade. During the 1955-1965 decade, aircraft controls had accounted for at least 50 percent of total sales and went as high as 70 percent in 1961. Meanwhile, hydraulic turbine controls were declining in relative importance, and engine and turbine controls were only slowly increasing in sales. In the 1966-1976 period, however, the trend reversed. Engine and turbine controls surged to become the company's product sales leader. In 1966, aircraft held a narrow 50 percent to 42 percent lead with hydraulic turbines trailing at 8 percent of total sales. By 1970, the Engine & Turbine Controls Division reached 50 percent for the first time and reached a high of 64 percent in 1975 before declining slightly to 60 percent in 1976.[10] The Fort Collins plant, headquarters for the Engine & Turbine Controls Division, had come of age during this time, but its efforts were aided by the financial success and rapidly increasing sales of the overseas branches—all of which emphasized engine and turbine controls. Aircraft controls dipped to 34 percent of total sales in 1972 and went as low as 31 percent in 1975 before rising to 35 percent in 1976.[11] Of course, the trend was for absolute sales to rise; thus, a division might sell more even as its percentage of total sales went down.

The company was conservatively and cautiously run throughout this decade. Irl served as brakeman to some innovations, but he was not opposed to every change. He proposed an expansion in the number of directors from five to seven. In December 1966, Walt Thorell and Dick Leighton were appointed to these openings.[12] Mary Barb Bittle replaced her husband Billy, who had resigned from the Board in May. Irl's good friends, Eugene Abegg and Fred Humphrey, and Elmer Woodward's nephew, George W. (Bill) Taylor were already on the Board. With the Board's support Irl continued to work beyond the company's mandatory retirement age of seventy. He justified this exception by proposing an amendment to the Profit Sharing Trust that excluded any active company member over seventy from participating. This singled out Irl, and for the rest of his life, he received no profit sharing.[13] He did continue to receive $2,000 a month for the next decade, a salary lower than that paid to numerous officers who worked for him. His considerable stock holdings, however, enabled him to lead a comfortable life. But, by contemporary standards, he was substantially underpaid as chairman. He had always said that work at Woodward was more important than the remuneration and he proved his point in this decade.

During this time, financing the costs of the company's expansion was a problem, particularly since Irl had always disliked the idea of putting the company into debt. The partial solution in 1966 was to use a portion of the members' Profit Sharing Trust to finance up to $1.5 million of the company's

Mary Barb Bittle

Cliff Olds instructs Colin Clark (left) and Hugh Hiigel, members of the first class of the Irl C. Martin Academy of Industrial Science.

building debt. In return, the company would pay interest to the trust at a slightly higher rate than it could get from other sources. Both the company and its members benefited from this plan.[14]

In 1967, three other innovations would also benefit the company. Discussion was begun on a long-term disability plan to be included as one of the tenets of the Corporate Partnership. This was promoted by Dick Leighton, who had commissioned an internal study to make recommendations. The directors liked the idea and decided that the insurance for this would be paid for out of the workers' share of Corporate Partnership profits—the normal way that additions to the Corporate Partnership were funded. The second innovation resulted from a study that Irl had commissioned from the company's Education Committee—one of the company's internal committees. The Education Committee had recommended that Woodward institute an academy to train future Woodward members. Irl asked the Board of Directors to adopt the committee's recommendation, which it did, and the Woodward Academy began in July 1967. The third, and most significant, was a plan to change the organization of the company from departments to divisions. The immediate result was to begin the process of moving all engine and turbine personnel to Fort Collins to occupy the new plant built in 1966.[15]

The purpose of the academy was to complement the seven-year-old, summer only, Recruit Training Program by offering high school graduates a two-year curriculum with equal amounts of work and study. A two-year graduate could work full-time with the company in a skilled position, find a job with another company, or pursue education full-time at a college or university under a leave of absence.[16]

The academy was a practical way to attract good people and train them in industrial and academic skills. Company leaders had anticipated that skilled personnel would be increasingly difficult to attract, and they hoped that students who had gone through the academy would either stay and become Woodward members or continue their education and then return to Woodward.[17] The first fourteen young men who completed the program in 1969 did stay with Woodward—nine stayed in Rockford and five went to Fort Collins. Each was credited with one year of seniority so that, by the end of the next year, he would get the full benefits of regular membership. The academic part of the 178-unit curriculum included mechanical drawing, algebra, physics, electricity, advanced math, management and other business subjects. Practical experience and more study followed on machine tools, characteristics of metals, power sources, tooling, and other machine shop basics.[18]

Irl C. Martin Academy of Industrial Science.

The school was "outstandingly successful" from the start. On December 12, 1970—Irl's seventy-fifth birthday—the Woodward Academy was renamed the Irl C. Martin Academy of Industrial Science to honor the man most responsible for the idea.[19] The honor was most appropriate since the Academy's program of half-day study and half-day work represented Irl's ideal learning situation. He had once attended college for two years himself, and although he had thought it a waste of his own time, he was now, quietly and without fanfare, financially aiding the higher education of other college students.[20]

After its first five years, Dick Leighton assessed the academy as a program "worthwhile for the student, the company and the community" and hoped it would be emulated by other companies. By this time, the academy had graduated thirty-five men in three classes. Only five had broken ties with Woodward, while twenty were working full-time for the company. Four had elected to continue their education, and the other six were on company leave while they were in the military service.[21] The academy has since continued as an integral part of the Corporate Partnership and remains successful in Rockford, and, since 1975, in Fort Collins. At first, in the 1960s, the academy's students were all male, but since then many women have been Academy graduates.

These three events of 1967, combined with the cost of the new plants in Fort Collins, the Netherlands, England, and Japan were a drain on the company's resources. Sales were up over $7 million in 1967, but profits were only $250,000 higher than 1966.[22]

Irl Martin and Dick Leighton were enthusiastic about the successes of the three overseas plants. All were profitable and each showed increasing sales. Overall sales increased $2.4 million in fiscal 1968, but profits of $2.2 million were practically the same as in 1967. Sales began to drop after 1968 and, in 1970, were over $2 million below the 1967 figure. Profits remained slightly over $2 million each year from 1967 through 1971. By most standards, these were successful although flat years, but there was dissatisfaction at Woodward. Dividends to stockholders, which had been $3.50 a

Each plant selected a member to present a gift to commemorate the company's 100th year anniversary in 1970. (From left) Nicol deKuiper (Hoofddorp) brought a piece of crystal; Pearl Whitfield (Slough) a Wedgwood plaque; and Kikutaro Haga (Japan) a stone lantern—along with a plaque bearing the names of all Japan members. President Dick Leighton (right) accepted the gifts on behalf of Rockford plant members.

share in 1965, were cut to $2 in 1966. They reached $3 in 1968 and $4 in 1970, but the Board of Directors still had to deal with stockholder complaints throughout this period about low dividends, demands for stock splits, and for listing the stock on an exchange. Worker members were also dissatisfied by the decline in profit sharing each year between 1966 and 1970.[23]

The company fought this dissatisfaction with an austerity program aimed at cutting the cost of doing business. However, necessary additional borrowing in 1967 and 1968 continued to slice profits. Finally, during fiscal year 1969, the company paid off all United States short term debt, and invested another $2 million.[24] Nevertheless, aircraft governor sales, primarily to General Electric, continued to decline and, when General Electric workers went on strike in 1969, it had a further negative effect on Woodward's shipments.[25]

The officers and directors had other worries beyond the economic health of the company during the late 1960s. Despite the stagnant profit picture, the company looked attractive to outside buyers. Their approaches concerned Woodward's management, which feared that a takeover battle, if lost, could mean the end of the Corporate Partnership. Accordingly, Leighton solicited advice from attorneys, public accountants, investment counselors, bankers, and officers from other companies on methods to forestall a tender offer. The plans included getting more worker members to buy stock and requesting IRS permission to invest up to 20 percent of the funds in the Profit Sharing Trust in company stock. Worker member stock purchase plans were discussed throughout 1969, but none was adopted. By the end of 1969, the company directors had discussed a series of precautionary procedures "to forestall a takeover if a tender [offer] ever materializes."[26]

Another issue that confronted the company during the late 1960s was minority employment. According to the Corporate Partnership philosophy, the company did not discriminate on the basis of race, creed, or color. It emphasized both equality and excellence. However, it also required extensive written tests as one condition for employment. In Leighton's words, "To many outsiders this appears to discriminate against the underprivileged who may not be prepared educationally or socially to take the tests." Although Leighton insisted that the Woodward hiring system was fair, he also noted the company's support for employee training, such as its own

Recruit Training Program and the local Opportunities Industrialization Center. These would help to make up for any deficiencies in education an applicant might have due to factors beyond his control.

In mid-1969, management was also under some pressure to do its part to help bring about better community relations in Rockford. In response, Leighton set up the Social Concerns Committee on July 31, 1969, as a temporary Vertical Committee of the company. While the committee had many general duties, its major responsibility was to find ways to increase minority employment within the company.[27]

Concerted efforts were made by management to comply with equal employment opportunity guidelines while educating individual members on the necessity of working openly and fairly with each other as equals toward a common goal. These steps put the company squarely behind its long-professed ideals, and the situation continued to improve as more minorities were hired throughout the 1970s.

Woodward celebrated its centennial on May 31, 1970—100 years from the day that Amos Woodward received his first water wheel governor patent. This centennial was highlighted by the presentation of gifts from Woodward's three overseas plants, by three open house events, and by the making of a company motion picture called "A Will and a Way."[28] Business improved by late summer, so a small amount of profit sharing was available to worker members at the end of the fiscal year in September.[29] A minor expansion also occurred in September 1970 when Woodward set up a branch office in Sydney, Australia.

Throughout 1970 and 1971, the company practiced austerity in a time when the pace of the United States economy was stagnant. The Vietnam War had helped unleash inflation, a continuing problem throughout the 1970s. The company was hit hard in 1971 when the supersonic transport (SST) project was cancelled by Congress. Termination claims were received, but they were a weak substitute for the business the SST project would have brought in. Fortunately, this loss of business was partially balanced when Woodward received contracts to provide the fuel controls for the B-1 bombers and for General Electric engines on the new McDonnell Douglas DC-10 passenger aircraft.[30] Nor did it help that a number of Woodward's customers were not paying their bills on time. To pay some of its

In 1970, after 100 years of operation, the company shipped its one millionth control (serial number 1,000,000). This hydraulic turbine governor for the U.S. Army Corps of Engineers was installed at the DeGray Reservoir Dam on the Caddo River in Arkansas.

Jerry Brannon (left) and John Spedding remove time clocks from the plant in January 1972. All members then went on a salary pay basis to be consistent with the principles of equality expressed in the Corporate Partnership philosophy.

own bills, the company had to redeem some invested securities. Irl was displeased in either case and suggested that, if customers did not pay according to the agreed-upon terms, Woodward ought to put them on a C.O.D. basis.[31]

In 1971, Irl provided another major contribution to the company—a written constitution, with a preamble much like the one in the United States Constitution. It was accepted by the Board of Directors on September 21, 1971, and affirmed by the membership on October 1, 1971. It would be Irl's living legacy. In it he charged his successors to support and defend the constitution and thereby keep the Corporate Partnership.[32]

Irl's contributions to company and member welfare continued in 1972. At his request, the Board of Directors had made two changes in the company bylaws in 1971. The first was to allow emeritus directors. At the January 1972 annual meeting, Dr. Fred Humphrey and Walt Thorell requested this status which allowed them to attend the board meetings and give counsel but not vote. The second change was to increase the number of directors from seven to nine. This allowed four new appointments, all company members: Bruce Livingston, vice president and general manager of the Aircraft Controls Division; Mark Leum, chief engineer and general manager of the Hydraulic Turbine Controls Division; Russ Martin, vice president and general manager of the Engine & Turbine Controls Division; and Paul Mann, director of Personnel and Public Relations.[33] The major purpose of this action was to bring younger men on the Board in order to take advantage of their points of view while training them for higher company offices.

A new member benefit Irl promoted in 1972 was a Retirement Income Plan as a supplement to the Deferred Profit Sharing Plan. This retirement plan replaced the pension plan in force since 1947 which had become inadequate, since it provided only $2 a month per year of service. The Retirement Income Plan, unlike the Deferred Profit Sharing Plan, was to be paid out only in the form of an annuity. This new benefit was approved by the Board of Directors on September 20, 1972.[34]

Another plan approved by the directors on that date was the Seniority Premium Plan. The purpose of this plan was to recognize the value of a member's continuity of service, thereby removing this element when determining an individual's base wage.

Business began to pick up in fiscal year 1972 domestically and overseas. Leighton described the mid-year results as "quite satisfactory." Woodward had, by competitive bid, beaten Hamilton Standard to supply the control for the new Harpoon Missile Project. The company's engineering group was enthusiastic about new product developments in all divisions, about the knowledge they acquired in electronics in developing the control for the Harpoon Missile, and about the design and development of a digital system controller. The digital system controller was seen as the "trend of the times," because it simplified controlling by using a mini-computer. The advantage was that the system could be changed by reprogramming, thus eliminating all the remachining and redesigning of the hardware.[35]

Fiscal year 1972 ended with billings of $43 million, up $6 million from 1971. This was the highest in the company's history, and new sales records were again set each year in 1973, 1974, and 1975. Profits followed the same pattern, more than doubling between 1972 and 1975.[36]

All divisions and all the overseas plants except the sales and service office in Australia were increasingly profitable in fiscal 1973, another record year. The largest increase in shipments was in the Aircraft Controls Division. This came about because of previous development efforts combined with a rapidly improving economy. Increased production quantities of 3034 and 3059 fuel controls for DC-10s were prime factors in sales and profits successes. However, the company also was producing the newly-developed main fuel control for the Navy S3A plane. Synchronizers for propeller-driven and jet aircraft also sold well. They were being put on all McDonnell Douglas DC-9s. Development projects at Fort Collins, which emphasized electro-hydraulic fuel controls for industrial gas turbine engines, were at an all-time high. Sales from the International Division were even going to the Soviet Union and other Eastern European countries, and Woodward governors on Rolls-Royce aircraft engines were permitted to be sold to the People's Republic of China.[37]

By the beginning of fiscal year 1974, the company had achieved such a complete turnaround from the 1970-1971 period that it could not find enough additional workers in the Aircraft Controls Division in Rockford. Members were asked to bring in friends and relatives for job interviews.[38] In Fort Collins, construction of a new administration building and an addition to the manufacturing plant continued during 1973 and 1974. This allowed all personnel to work at the plant complex on Drake Road, and the downtown buildings were sold.

Both plant and member expansion occurred during 1974 to meet the increasing demands for Woodward products. Shipments were generally higher than projections, but costs were also rising rapidly, especially in parts furnished by suppliers. Scrap costs were also going up, mainly due to the process of orienting new members. Overtime hours were increasing and bonus prospects were good. Woodward had sufficient internal funds to provide for capital expenditures, such as the final construction costs for the new buildings in Fort Collins and for the purchase of land for the projected plant in Tomisato, Japan.[39]

The most promising product line was the 3013 fuel control for the new Fairchild Republic A-10 close support, attack aircraft. *Prime Mover Control* carried an enthusiastic article about this aircraft in mid-1973, well before it went into production. By early 1974, the company projected a 50 percent increase in its Aircraft Division by 1978 if the A-10 won "a fly-off" against the rival A-7D. The A-10 won and by the end of 1974 the first sixty A-10 controls were in production. This product continued to be important in 1975 and 1976. However, because deliveries were stretched out, it didn't reach the sales projections that the company made in 1974.[40]

Aircraft Controls remained the glamour division between 1974 and 1976. It was glamorous because of its close ties with military projects like the A-10 fighter and B-1 bomber projects, but also because aviation was the place where the most exotic and sophisticated technological research was happening. However, the Engine & Turbine Controls Division was providing well over one-half the company revenues in the 1971-1976 period without having any glamorous products. This division benefited from the rapidly rising oil prices that followed the 1973 Arab oil embargo. This led to greatly intensified exploration activity in the oil and gas industry. The sale of Woodward controls thus substantially increased for oil and gas industry application. For example, the oil crisis led to a "substantial number of orders" for the new 43027 industrial electronic gas turbine fuel control. The sale of hydraulic mechanical controls for diesel generating sets also increased substantially. Sales at the Hydraulic Turbine Controls (Hydro) Division in Rockford continued to grow from 1973 through 1976. Because of the growth in other divisions, Hydro continued to be responsible for only 5 to 6 percent of total company sales each year. Still, by 1976, the level of unfilled orders for this division was at an all-time high.[41]

As the company's profits increased in the mid-1970s so did dividends and the profit sharing percentage. Dividends reached an all-time high in 1976 at $6.75 per share, and the cash profit sharing was over 52 percent. Another member benefit was also added during this time. The company had a dental plan since 1957, but the program had not included members' families. In mid-1974 a new plan was proposed to include spouses and dependents. During 1975, it became part of the new Health Care Plan that was approved by the membership and the Board of Directors. The plan went into effect on October 1, 1975.[42]

By autumn 1975, Irl Martin's health had declined so badly that Cal Covert, who was in Japan on company business at the time of the September Board of Directors' meeting, returned as vice chairman of the company. Irl had promoted a change in company bylaws and nominated Cal. The board unanimously approved the appointment.[43] It was clear that Cal was Irl's heir apparent. In the next six months, Cal and Irl were almost inseparable as Cal tried to learn everything about how Irl thought the company should be run.

When the annual meeting of the Board of Directors was held in Rockford on January 7, 1976, Irl was too ill to attend. After some of the topics were discussed, the directors decided, as a measure of respect to Irl, that the meeting should adjourn and reconvene at Irl's home in Phoenix, Arizona, on January 15. There, primarily for Irl's benefit, they reviewed decisions made in Rockford before deciding a few other minor issues.[44]

Irl's health rapidly worsened in the next two months, and he officially resigned as chairman of the board on March 24. His short resignation letter thanked all company members for their "loyal support," and he challenged them to preserve the Corporate Partnership "by the best efforts of your time, talent, energy, and will-to-do."[45] An era that had begun with Irl's employment in 1921 had ended.

The Board of Directors met in a special session in Scottsdale, Arizona, and accepted his resignation on the day it was tendered. As part of the motion of acceptance, the Board of Directors expressed "its deep appreciation" to Irl for his more than fifty-five years of service to the company. The resolution continued:

From left: Dorothy Martin, Cal Covert and Irl Martin.

Irl's understanding of his fellow members, belief in their integrity and concern for their well-being were significant factors in building a working environment unique in the world's business community... Individual dignity and quality of product always came first with Irl. In a very real sense the Woodward Governor Company today is a product of Irl Martin's philosophy, ability to communicate and hard work. Irl will be sorely missed but not forgotten...[46]

Director Russell Martin then nominated Cal Covert as chairman, and Irl's daughter, Mary Barb Bittle, seconded the nomination, which was passed unanimously. President Bruce Livingston moved to appoint Irl as chairman emeritus, and Bill Taylor seconded the motion.

Irl receives his pin for 50 years of service.

Another unanimous vote followed. Cal's first statement as chairman was to pledge to "continue the work that Irl had initiated" and continue "those policies and principles established by Irl."[47]

Within a month, on April 22, 1976, Irl died from cancer and heart disease. Cal had been with him almost constantly that last month, during which Irl continued to tutor Cal on the subject of corporate leadership. The day after Irl's death a memorial service was held for him, appropriately, on the production floor of the plant in Rockford. He was eulogized by the company chaplain, Reverend Ralph Fry, for the quality of his life and for all he had done for others. Fry concluded, "a great man has fallen this day."[48] Rockford radio station, WYFE, eulogized him, saying:

The Rockford community will never be quite the same with the passing of Irl C. Martin, eighty, an industrialist of the type we may never see again… Woodward has been unique—not only for the superb techniques applied to the application of hydraulic controls—but also for the indelible stamp Martin placed on staff and operations.[49]

Practical to the last, Irl donated his body to the University of Illinois Medical School. He thought that doctors might learn something from the condition of his hip sockets which had been replaced at a time when the operation was uncommon.[50]

With Irl's death, Cal Covert was fully in charge of the corporation. He had attained the highest position in the company without actively desiring it; director of facilities planning had been his highest aspiration. However, Irl's influence and friendship and Cal's abilities led to more promotions which forced Cal to broaden his knowledge of the company and its operations. How he would handle the problems almost certain to come would be his major task in the coming months and years. He must have felt the same sense of awe Harry Truman described to newsmen in 1945, when he said that he felt as if "the moon, the stars, and all the planets" had fallen on him after Roosevelt died.

A Time of Growth

1976-1989

Cal Covert faced all the usual challenges that confront every new corporate chairman. He had products to sell and a profit to make, new markets and technologies to research, strategies to plan and execute, customers to serve, and stockholders and fellow worker-members to satisfy. Fortunately, the company had a strong tradition and loyal work force to help him meet these challenges. The tradition had been set in place by Irl, who had once told Cal, "The people make the company."[1] The "people," in this case, was not an abstract entity of names and statistics filed away in a cabinet at Personnel. They were real human beings and Cal recognized, as Irl had before him, that his primary responsibility was to them.

Some challenges, however, were not so usual. Beyond the company's doorstep, the economy was in disarray. Inflation, combined with low growth and high unemployment, choked economic activity. Industrial productivity increases during most of the Carter administration averaged only about 1 percent a year, a mere quarter of the average increases in West Germany and Japan. Accordingly, the standard of living in the United States declined from first to fifth in the world. High interest rates bordered on the absurd. Real income dropped two percent a year from 1973 through 1981.[2] Foreign-made products, once the butt of jokes, were flooding into the American kitchen, living room, and driveway.

To many people in the 1970s, it seemed as if American society as a whole was in decline. Many writers, journalists, and politicians recorded as a matter of fact America's decline as a world economic and military power. President Jimmy Carter, only a year into office, spoke to the nation of a "national malaise." Americans, it seemed, were becoming lazy, their work performance was shoddy, and their commitment to the values and ideals of the American experiment in liberty had declined. It was the decade writer Tom Wolfe called the "me decade." In place of the sometimes misguided idealism of the sixties, people began to turn inward to seek lifestyles of personal enrichment and comfort.

Calvin C. Covert, Woodward Chairman and Chief Executive Officer

Much of the "me" generation's lifestyle might be looked upon as a refinement of the "practical philosophy" Tocqueville saw in America in the 1830s. Men and women were enriching America, and themselves, under the category of getting things done. But there was a darker side to it as well. For one thing, men or women might too narrowly focus their loyalty on their own desires. Loyalty to a company, a profession, or a community would be secondary to loyalty to one's self.

This spirit of the times was also exerting pressure in American business. The seventies and eighties saw the rise of corporate raiders, leveraged buyouts, junk bonds, and arbitrageurs—a whole range of financial wheeling and dealing where the primary motive was to make money, and lots of it, by buying and selling companies instead of the products and services they offered. The new business climate was potentially devastating to the traditional values found at Woodward. If managers, after all, were no longer loyal to a company and its people, who could expect loyalty from the employee? As a consequence, many people no longer saw themselves as working for a company for life, but as toiling in a profession for the highest bidder. Families would uproot themselves from one state to another to go where better and higher-paying work beckoned. College students enrolled in business and professional schools, because the best-paying jobs awaited those with business degrees. By the same token, companies were unceremoniously letting long-time loyal employees go, because strategies and owners had changed. Loyalty, often described as a two-way street, had become a one-way detour pitted with holes and ruts.

Nothing could be more contrary to the Woodward Way than the worst aspects of the "me first" attitude described above. Thus, Cal faced, above all other challenges, one that confronted no other corporate chairman, and that was to protect and strengthen the Woodward Way.

Cal Covert believed in the Woodward Way. For him it was the program for an ethical way of life, as well as the proper way to run a business. He was born in Rockford on November 24, 1924 and, except for a three-year stint in the Army Air Corps during World War II, had lived all his life in Rockford. He was educated in the Rockford school system. By the time he was eleven, he was working at Edgebrook Golf Club as a groundskeeper, where his interest in landscaping and horticulture developed. Through occasional caddying, he became acquainted with Elmer Woodward and George Sorensen. They suggested that he put in his application at Woodward. By the summer of 1942, Cal was essentially in charge of grounds at the golf course, when he decided to act on this advice. On June 13, 1942, Cal was hired by Woodward to work on grounds maintenance at the new North Second Street facility. The next May, Cal was transferred to Diesel Assembly but, at the same time, he felt that his "place in the world was to be patriotic," and he volunteered for the draft.[3]

He was trained as a radio operator and served with the Air Force in the China-Burma-India theater. He left the Army Air Corps in March of 1946, and immediately returned to Woodward. His new duties put him to work at one of the toughest jobs in a factory. It was called "snagging," the deburring of the insides of metal castings. After two-and-a-half months of this, he went back into grounds maintenance and progressed from yard worker to chief yardman in 1949, and later to supervisor. All during this time, he studied through a correspondence school in Des Moines, Iowa, courses related to grounds management and physical facilities. He was appointed corporate manager of Facilities Planning in the 1960s, which involved him in the construction of new plants in Fort Collins, Japan, and the Netherlands. On December 18, 1970, he became a company officer when he was named assistant secretary. In January 1973, he was voted to a place on the Board of Directors, and two years later was named vice chairman of the board and secretary of the company. He was appointed chairman on March 24, 1976.[4]

Cal's relationship with Irl began the day he reported back to Woodward after the war. Although federal law required the rehiring of returning war veterans, companies were under no obligation to place them in their former positions. Cal thus found himself working inside the plant, in the Snagging Department, since that was where help was needed. He would have left Woodward had it not been for the encouragement of a friend with whom he worked, and the fact that Billy Bittle, Irl's son-in-law, was also snagging. Cal snagged steadily, day after day,

Cal Covert working in flower bed – 1948.

until on one of those days he was approached by Irl, who said, "I see you stuck it out. We admire you for that. How would you like to transfer?"[5]

Once back outside and working on the company grounds, Irl and Cal grew closer, eventually forming a father-son bond that lasted until Irl's death. A part of Cal's responsibility was to tend the grounds at Irl's home (and Billy Bittle's, next door) and to help him with remodeling and building projects. The future company chairman thus learned construction skills to supplement his skills in landscaping. These skills would prove helpful to Cal in his subsequent position managing all new Woodward construction projects.

Such closeness between the two inspired more than a technical education. Irl's commitment to the Woodward Way and Corporate Partnership was infectious, and Cal found himself enthusiastically in support of it.

Usually it was the chief subject of conversation between the two, especially when they traveled together to the construction sites. During the early phases of the Fort Collins project, they shared a room. Like master and student, they talked at length about the importance of the Corporate Partnership to the company. And it was not ideals alone they talked about, but the deeper values of fairness and a working man's dignity. To Irl, the Woodward Way was an entire philosophy of life.

Cal learned that a responsible business was more than a company that made a good product and sold it at a fair price; it set the standards for ethical living among its employees and in the community at large. At the early recruit classes, for example, young students were taught about the company's products and its history. They were taught basic economics and some practical politics. But they were instructed also in such basic life skills as social morality and self-discipline. A chaplain spoke to them about the emotions of growing up. It was Irl's plan that the young recruits become good citizens and not merely good workers.

Cal accepts award from Rockford Magazine *as Business Leader of the Year – 1988.*

None of this was lost on Cal, who took to heart perhaps the most important lesson his mentor had to teach. A company is made up of people and, unless a manager knows how to treat people with fairness and dignity, he is no manager at all. Cal thus came well-prepared for the challenges of corporate leadership. Besides his wealth of experience with the company, he had considerable common sense which was valuable in his day-to-day contact with fellow members. His highest word of praise for another person is to call him or her a gentleman or lady, which signified one's concern, commitment, and fairness. Standing above all this was his commitment to the tradition begun by Irl Martin. It directed his goals for the company's business picture, and guided his philosophy of care and self-discipline with company members. "All I've done," he once said in a magazine interview, "is capitalize on the foundation Irl built."[6] The modesty of this statement cannot obscure his influence or success. He was named by *Rockford Magazine* as the city's business leader for 1988.

One challenge Cal and the company had to take on was the problem of presidential succession. In the early 1980s, Bruce Livingston, who had been appointed president by Irl in 1972, was stricken with heart disease. He underwent heart valve surgery, which required a long period of recuperation away from work. His illness thus exposed a weakness in the Corporate Partnership Cal would have to correct. The partnership stressed the importance of promotion from within the company but, at the time of Bruce's illness, no one was perceived as being ready to take his place. After Bruce returned, he was asked by Cal and the Board of Directors to institute a management succession program to prepare members for promotion to the highest positions in the company. Bruce barely had time to get started, however, when his health declined again, this time to the ravages of cancer. Unable to continue on, he retired in September 1982, and died July 9, 1984.

Cal Covert and Bruce Livingston.

The absence of a Woodward member prepared to assume the position of president forced the company to make a controversial decision. It was decided that a search for a president from outside the company would be made, and management consultants were hired to help screen candidates.[7] In the meantime, Cal filled the position as president. After a six-month search, the presidency was offered to Robert D. Pope on March 5, 1983. He accepted and joined the company on April 5.[8]

Bob Pope and Cal Covert.

Pope was dynamic, decisive, and strong-willed, a professional manager determined to take charge and institute far-reaching changes. He urged changes in the member evaluation system to emphasize accountability for performance and he, likewise, placed greater emphasis on financial planning and discipline. He was confident, articulate and set high standards, but could also be intimidating. In time, it became clear that he and Cal simply did not see eye-to-eye on how to run the business. By 1987, he had lost the support of the board, and he resigned on March 25.

By now, though, the company was prepared to promote from within, and to fill the position of president, Cal chose Mark Leum.[9] Mark had joined Woodward in 1952 as a mathematician working with the development engineering group. He had been involved in all the company's product lines and, for twenty years, had been general manager of the Hydraulic Turbine Controls Division. He had also been a member of the Board of Directors since 1972. His breadth of experience made him the typical Woodward man, and he was strongly committed to

President Mark Leum (right) and C. Phillip Turner, then corporate secretary, confer.

the Corporate Partnership. Once he became president, the company's internal leadership strife came to an end.

Another challenge facing the company was the Corporate Partnership itself. The company was no longer run by blood or marital relatives of the Woodward family, but it was still a "family," even though its members were scattered all over the world. Keeping the family intact, despite the great geographical distances between them, meant keeping the Corporate Partnership alive. To do this, the company would have to remain independent and in control of its own fate. If it were to merge with another company, the first thing likely to go would be the Corporate Partnership.

The process was carried on through gradual reorganizations of the structure and makeup of the board, and by amendments to the company's bylaws and articles of incorporation. In 1977, Vern Cassens, company treasurer, was appointed to the board, as was William O. Nelson, executive vice president of the J. L. Clark Manufacturing Company of Rockford. In 1979, Jack Kasper, chairman of the board of Elco Industries, replaced Bill Taylor. In 1981, J. Peter Jeffrey, chairman and president of Illinois National Bank and Trust Company, replaced Irl's old banking friend, Eugene Abegg, who had died in 1980. These men joined a board already made up of Mary Barb Bittle, Mark Leum, C. Russell Martin, Bruce Livingston (until his retirement in 1982), and Cal Covert. They provided the support Cal would need for keeping the company's independence intact.

Even before the board was reorganized, it was active in strengthening the Corporate Partnership. On September 21, 1976, only weeks before Cal became chairman, it approved the twenty-seventh amendment to the Deferred Profit Sharing Plan. Under this amendment, the trustees of the plan would be allowed to invest 100 percent of the trust's assets in Woodward stock. But a resolution was also passed requiring a referendum of worker-member participants before any substantial disposition of the stock could be made.[10] The amendment, therefore, would allow the plan to buy stock that might be the target of a leveraged buyout, but the referendum also preserved the members' right to approve the purchase. It was a wise amendment. Woodward stock increased in value, and the assets of the trust brought benefits to individual participants. By 1987, about 23 percent of the company's stock was owned by worker members through this trust and through their personal stock holdings.[11]

Other changes continued. On January 10, 1977, the company was reincorporated in Delaware. In June 1980, when the over-the-counter price of Woodward stock was about $380 a share, the board declared a four-for-one split to "insure the maintenance and continuation of a reasonable market" for the stock. This decision was approved by the stockholders at the January 7, 1981 annual meeting. Authorized stock was increased from 240,000 to 960,000 shares.[12]

A more important proposal was also approved at the 1981 annual meeting. It required that the Board of Directors "shall be guided by the philosophy and concepts of human and industrial association of the Corporation as expressed in its Constitution."[13] This was the same constitution Irl had drawn up years before, in which he had expressed the values of human fairness in the practice of business. The resolution passed at the meeting did more than call upon the board to support the Corporate Partnership, however. It explicitly reinforced the philosophy as a partnership among all shareholder and worker members of Woodward.

The most important measures that were passed at the January 11, 1984 annual meeting of shareholders included a two-for-one stock split and an increase in the authorized shares to 5 million. This meant that slightly under 1.5 million shares would be outstanding with 3,480,000 shares authorized and available for issuance. The additional shares could be used for a variety of purposes including financings, worker member benefit plans, stock splits, and stock dividends, but the proxy statement was explicit in noting "The Company could use additional shares of the Common Stock to create voting impediments or to frustrate persons seeking to effect a takeover or otherwise gain control of the Company..."[14]

Further changes to the company's bylaws were made at the 1984 annual meeting. The board was given more power, but a measure was also passed providing for concurrent authority by a two-thirds approval of the outstanding shares to amend company bylaws. The board was divided into three groups of three members, each group serving staggered three-year terms. Also, a board member could only be removed by shareholders for cause. Finally, a change was made to the procedure for holding special meetings of shareholders. Only a 20 percent vote of shareholders had been required to call a special meeting. Now it would take a two-thirds majority.

Aside from changes in corporate structure and organization, protecting the Corporate Partnership also meant looking after the more immediate interests of the worker-members. In pursuing this, the company had upgraded its membership insurance program, encouraged members to purchase and retain company stock from their private funds, and also allowed members to use portions of their deferred profit sharing funds to buy stock. Advisors for personal finances and taxes, insurance, and legal problems were also provided and paid for by the company under terms of the Corporate Partnership. Most company members were, therefore, able to retire comfortably by age sixty-two. Some retired members found that their deferred benefits appreciated in value each year by more than the amount they took out to pay for living expenses.

A testimony to Woodward's concern for its members was given by *Forbes* magazine in 1986. Robert McGough, the reporter who wrote the story, noted the significance of the induction ceremony which was held for members who successfully completed a two-year probationary period. "Imagery and ceremony play large roles. To work at Woodward you don't merely sign a time card and show up for work. You consciously and solemnly join a corporate family."[15]

In recent years, first-time visitors to Woodward plants often compared the company's philosophy to the Japanese way of doing business. The Japanese way, which burst upon the American consciousness in the 1970s, was called "Theory Z" by University of Chicago business professor Willard Zangwill. Among other things, Theory Z identified "lifetime employment, collective responsibility, an overall concern for the worker" as reasons to explain why Japanese businesses were surpassing American businesses. What surprised many visitors, however, was to learn that these same principles were built into the Woodward Governor Company Corporate Partnership before Japan had recovered from World War II, before such successful companies as Toyota, Mitsubishi, and Sony had poured the foundations of their new post-war factories. No one at Woodward knew for certain if post-war Japanese business leaders were aware of the Corporate Partnership, or if they were copying and adapting parts of Woodward's business philosophy. Professor Zangwill, however, agreed "it is conceivable that [Woodward] was a model for Theory Z."[16]

Since 1981, the company directors had been charged with upholding the Corporate Partnership philosophy expressed in the company's constitution; they are legally bound to protect it.[17] Protecting it was thus Cal Covert's highest priority. As it had evolved since its beginnings in 1946, it had continued to benefit worker and shareholder members alike. Irl Martin wanted the Corporate Partnership to survive because of these reasons and, under Cal's leadership, it was as strong as ever.

While Cal consolidated his position as chairman and strengthened Woodward's Corporate Partnership, the company's business and financial picture continued to improve. During the period of 1972-1982, encompassing Bruce Livingston's presidency, net billings increased from just under $44 million to almost $159 million. Net earnings per share likewise increased from $3.36 to $17.55. The shareholders' equity went up every year, nearly tripling over the decade, as did dividends. Worker members saw regular increases in their wages, in part due to high inflation, and cash profit sharing that sometimes exceeded 50 percent even before deferred profit sharing was distributed.[18] The first drop in net billings, $3.5 million, came in 1983 but, since then, they accelerated, reaching $299 million in 1989.[19]

The company's success was due to its products and future potential. The divisional structure remained unchanged until 1989, although the relative economic importance of each changed over the years. The Engine & Turbine Controls Division began to outpace the Aircraft Controls Division in dollar volume as early as 1970. From then through 1982, it accounted for more than 50 percent of total annual sales. In fact, it had 60 percent or more in 1975 and 1976, and nearly that much in 1977 and 1978. Then, sales in the Aircraft Controls Division expanded in 1983, surpassing 50 percent in 1985 and approaching 60 percent in 1986.

By 1977, Woodward's strong relationship with General Electric, dating back to the mid-1950s, became even stronger. G.E.'s CF6-50 series of jet engines had become the engine of choice for most commercial airliners, including some Boeing 747s. This engine was equipped with Woodward's 3059 control, orders for which doubled in 1977. The G.E. TF34 engine used on the Air Force's A10 close-support fighter was also equipped with the Woodward 3013 control, which was produced in large numbers beginning in 1978.[20]

Military applications received a minor setback in 1977, however, when the Carter administration cancelled the B-1A bomber program. The B-1A was to have been equipped with Woodward's 3055 control, so its cancellation was a disappointment. When President Reagan revived the program in 1981, however, production of the B-1B began, and so too did production of the 3510 control. In the meantime, the cruise missile, with Woodward's 3209 control, continued in production. The 3209 control allowed Woodward to establish a close working relationship with the Williams Research Company, manufacturers of the engine for the missile.[21]

During this time, the company continued to make governors, synchronizers, and synchrophasers for propeller-driven aircraft. Woodward's 3212 control for turboprop engines completed design assurance tests in 1978 and preproduction units were shipped to Pratt & Whitney of Canada. By the next year, Woodward controls were on the DeHavilland DASH 7 short-haul passenger planes. Woodward had also been chosen by a new customer, Rolls-Royce, to provide a fuel control for their engines which were to power the new Boeing 757 and 767 airliners. Finally, General Electric's joint venture with SNECMA of France (CFMI) led to the production of an engine used to replace older engines on the DC-8 airliner. The old engines were no longer fuel efficient, but the new ones with Woodward's 3556 control were.[22]

Re-engining of the DC-8 began in 1981, and CFM-56 engines were needed for flight certification tests in 1982. Woodward supplied the controls. Also in 1981, Woodward engineers began development of a full authority digital electronic control (FADEC) for aircraft turbine engines. Actually, the Engine & Turbine Controls Division had already developed digital controls for some of its applications. FADEC promised to be the company's next major breakthrough—digital controls for aircraft turbine engines. Between 1981 and 1987, Woodward continued experiments with FADEC because these sophisticated controls had the potential to provide "engine performance capability unachievable from a completely hydromechanical system." Woodward began making proposals for FADEC-based controls in 1983. Its value to the airliner market was demonstrated in 1986, when G.E.'s own FADEC was used on its most popular and successful gas turbine engine, the CFM56-5. Woodward, by this time, had developed the complementary hydromechanical unit (HMU) to translate the commands of the digital electronic control into the mechanical movement of engine parts. Its first application was on the Airbus Industrie A320 aircraft.[23]

Another product with apparent high potential was the company's 3278 fuel control for the G.E. engines on the General Dynamics Air Force F-16 fighter. These went into production in 1984, and by 1986 these same controls were on the Navy's F-14 fighter as well. They helped the Aircraft Controls Division achieve rapidly increased sales for military applications.[24]

Although the Engine & Turbine Controls Division, headquartered in Fort Collins, had fallen behind the Aircraft Controls Division in terms of total sales, its volume had, nonetheless, increased over the years. In 1976, sales were $46 million; this figure nearly doubled to $81 million in 1986 and, in 1989, this division's sales were $125 million. During the division's rapid expansion between 1971 and 1981, the dollar value of shipments increased at an average annual rate of 15 percent. Energy-related markets were the principal stimulus. The worldwide rise in oil and natural gas prices encouraged a demand for Woodward's products used for oil and gas exploration, production, and pipelines. At the same time, the need for organizations to make the most economical use of their energy requirements during a time of rapidly-increasing energy costs, caused a considerable demand for the company's sophisticated electronic controls. Sales of this division's electronic controls increased much more rapidly than the hydromechanical controls.[25]

By the early 1980s, the prices of oil and gas were beginning to flatten out, inflation was being brought under control, and the recession was coming to an end. These events especially aided the Aircraft Controls Division, but they caused the Engine & Turbine Controls Division to change its approach in seeking customers and meeting their needs. The main shift was from an emphasis on selling products to original equipment manufacturers, to selling to a more diversified group of end users. Consequently, market efforts were increased and field sales offices expanded. Authorized Independent Service Facilities (AISFs), which Woodward had begun in 1971, continued to grow in number. These strategically located service centers are independently owned but provide Woodward-approved service. By 1990, there were twenty-four AISFs throughout the world, each with overhaul and test capabilities and some selling new controls.

The Hydraulic Turbine Controls Division was a steady performer throughout most of the 1976-1987 period. It never failed to account for less than 6 to 8 percent of total company sales. The division passed the $10 million sales mark in 1982 and reached a record high in 1984, before dropping to $6 million in 1986.[26] The decline was anticipated, however, and the company was able to use the excess production capacity within the division to put members to work on aircraft controls. A gradual recovery was expected, and new products were likely to aid the recovery. One promising product was the result of a joint venture between the Engine & Turbine Controls and the Hydraulic Turbine Controls divisions. This was the menu configurable (MC) digital control for small- and medium-sized industrial turbines. It used elements from the Woodward 505 digital control for steam engines and applied it to control a hydraulic turbine. Its major purpose was to promote a more stable speed control for the turbine. It was especially attractive because it could replace the mechanical controls. The first unit was put into service in 1987.[27]

Woodward Chairman Cal Covert (right) joins Japanese members in raising the flags of Japan and the United States to begin the formal acceptance ceremonies of the new Tomisato facility on March 15, 1978.

To keep up with its expanding business, the company built new plants and added to others during this period. In 1976 and 1979, additions were made to the machine and assembly areas in Rockford. In 1977, construction began on the new plant at Tomisato, Japan. When it was completed in 1978, all operations moved from Tokyo. By this time, a small operation, primarily a sales and repair facility, had also been opened in Campinas, Brazil. The Australia operation, located in a suburb of Sydney, was expanded three times—in 1976, 1978, and 1979. Expansion also took place at Fort Collins, including additions in 1979, 1981, and 1989.

The major expansion project of the eighties occurred in Stevens Point, Wisconsin. It began with the decision in 1981 to lease a temporary facility. In 1985, construction was begun on a 136,000 square-foot plant, which was occupied in 1986. In 1989, a 47,600 square-foot addition was completed. This building was the home of the Hydraulic Turbine Controls Division and the newly-established Test Equipment business; it was also a major supplier of parts for the Rockford-based Aircraft Controls Division.[28] In early 1989, a 35,000 square-foot plant (plus full basement) was completed in Rockton, Illinois, to house the new Aircraft Product Service Center.

When we remember that Amos Woodward began by building a water wheel governor with a response time measured in minutes and that, for the rest of his long life, water wheel governors were all the company made, we can begin to appreciate how much Woodward Governor Company had changed in 120 years. The difference between mechanical technology in the 1870s, and high tech electronic and electro-mechanical technology of the 1980s seems greater than the difference between the meanings of the words "crude" and "sophisticated." The world of aviation, advanced energy systems, tiny microchips with storage capacity surpassing that of the largest libraries, and automated, robotic manufacturing systems hardly seemed to exist on the same planet where Amos made his contribution to the water wheel industry. But it did. The same company which made thirteen water wheel governors in 1871 now made controls for many of the world's advanced aircraft.

The world of tomorrow at Woodward looked even better, as the company continued to meet the challenges facing it. The Corporate Partnership became even more secure as the company shifted its product structure from divisions to groups–Industrial and Aircraft–and each group was further structured by business units. Justification for this shift was given in the company's mission statement, printed in the 1988 annual report:

We are committed to becoming a market needs driven organization serving prime mover control and accessory markets. Our continuing objective is to thoroughly understand and be responsive to the needs of these markets. We will continually search for new and emerging market needs while striving to provide our customers with greater value through new and improved products and services. We will continue to perpetuate and strengthen the Corporate Partnership philosophy.[29]

The announcement to structure the company by product groups was made by Cal Covert at the January 11, 1989 annual meeting of shareholders. He began by recalling the company's decision, made more than forty years earlier, that a "Woodward plant...should be no larger than 500 members."[30] The advantages of this figure were already proven in the company's recent history. Five hundred was about as large as a single plant could be and still expect to maintain effective communication. People got to know and work with each other, making for a friendlier working environment and a superior product. Membership evaluation was made easier, too, because people on the committee had "some personal knowledge of every member [they] evaluated."[31] Also, the 500 maximum made it easier for managers, since open communication and personal knowledge among members was vital to Woodward's vision of a harmonious, productive work force.

The limit imposed upon plant site numbers also fit in perfectly with the company's belief in fairness and success in the marketplace. Too many people would make communication difficult, and the information needed to run a business on ethical principles could become garbled. Cal referred to this in his speech: "The fostering of fundamental values such as God, country, and family will enable these people to function best in business and in society in general. The ability of people to communicate and work in such an atmosphere creates the foundation for future expansion and growth."[32]

In actual practice, structuring the company into product groups did not mean wholesale changes to its basic organizational structure. The Hydraulic Turbine Controls business unit, in Stevens Point, had fewer than 100 members, and the company had plans, said Cal, "to locate another business unit [Test Equipment] in the Stevens Point plant."[33] The Engine & Turbine Controls Division in 1989 was changed to three units: Reciprocating Engine Products, Rotating Turbine Products, and Plant Services. Also, the company divided the production of new aircraft controls into units for large and small gas turbines. Each unit, in keeping with the company's mission statement, was created to respond to new and existing conditions in the prime mover control and accessory markets. This would allow Woodward to identify more quickly the directions these markets took, and develop the products and services these markets needed.

In keeping with its plans to organize by business units, the company moved the Aircraft Product Service Center (APSC) to Rockton, Illinois, just a few miles from the Rockford plant. The APSC was originally established in the Rockford plant in 1987, and was transferred to the Rockton site in early 1989. By September 1989, the APSC employed 170 members and had sales of $39.8 million for that fiscal year. Its primary function was to repair and overhaul Woodward aircraft controls. Through its parts serial number tracking system, customer representatives and Woodward technicians could track a control through the repair process and perform the necessary overhaul in reduced turnaround time. It could also respond quickly to emergency needs, as when an aircraft is grounded.

The APSC came about as a result of the expansion in commercial aviation during the early and mid-1980s. Older aircraft, such as the McDonnell Douglas DC-8, were

Woodward's Stevens Point, Wisconsin, plant

operators had been enthusiastic. They especially appreciated the quality control and rapid turnaround time, which would have been difficult under the old system, when repair and overhaul was performed in the same organization having responsibility for production of new controls.[34]

Woodward's sales of aircraft controls increased significantly enough that the company saw a need to put its own test equipment on the market. In the past, Woodward had built test stands for its own internal use and for those few customers who preferred to repair Woodward controls themselves. But, with Woodward controls installed on almost every kind of aircraft there was, selling test equipment would meet what the company perceived to be an important need in the aviation community. Test equipment was being built at Stevens Point, since it was similar in size and construction to the Hydraulic Turbine Controls' cabinet actuators. Manufacturing facilities there were also easily adaptable to test equipment, as were the technical skills of the members. The Stevens Point plant also had the room to expand. Building test equipment for controls built by other companies, and even for equipment outside the governor field, was also under consideration.

A look at the company's newest products, and those in planning, illustrated the technical variety of its products. Its Phase III fault-tolerant digital control system (DCS) used microprocessors programmed "to control or sequence specific functions for many different applications."[35] Woodward's DCS was used, for example, at the

being refitted with new engines that used more advanced Woodward controls. Other established aircraft, like the Boeing 737 and 747, were being reconfigured, made larger, or equipped with newer engines, again supplied with Woodward's newer controls. There were also new aircraft coming into production: the Boeing 757 and 767, the McDonnell Douglas MD-80 (a derivative of the DC-9), and the Airbus Industrie A320. With new equipment also being designed for aircraft under development, it became apparent that Woodward needed to create a separate and independent business to see to the repair and overhaul of its controls. Since an increasing number of Woodward fuel controls were flying on European airlines, an overhaul facility had been added to the Hoofddorp plant. The response to the APSC from commercial aircraft manufacturers and

Reedy Creek Energy Center, which supplies electricity and steam for the Disney World theme park in Florida. Reedy Creek uses the G.E. LM5000 turbine, an industrial version of its CF6-50 aircraft engine. The Woodward DCS controlled and governed the speed and sequencing of the LM5000 turbines. The DCS was also coupled to the IHI-Sulzer 7RTA84M engine which powered the super-tanker Cosmo Venus. This application of the DCS was specially adapted to slow-speed engines, and prevented the engine from overspeeding and causing excessive vibration.[36]

New hydromechanical controls, which traced their lineage back to Amos's original water wheel governor, represented another area where Woodward responded to customer needs. Perhaps the two most important applications were on G.E.'s T700 engine for the Sikorsky Blackhawk helicopter and its unducted fan engine for commercial aircraft. The T700 powered the Army's UH-60A and UH-64A Apache and the Navy's SH-60B attack helicopters. In addition, a commercial version of the engine, the CT7, powered the Saab 340 and CN235 commuter aircraft, each of which used Woodward controls. The unducted fan engine was a project begun by G.E. several years ago to produce a new kind of fuel-efficient engine for aircraft. It is immediately recognizable by its turbine blades, which are installed outside the main engine duct rather than inside. The engine resembles a traditional propeller engine, with the blades at the back rather than the front. For this engine Woodward provided a pitch drive unit to regulate the pitch of the blades.[37]

All this activity required more than good engineering and sound marketing strategy. How a company manufactures its products is an equally important clue to its commitment to customer needs. The 1988 annual issue of *Prime Mover Control* highlighted this belief: "To improve our manufacturing procedures, we combine new techniques and procedures with those that are proven, for efficient ways to machine quality parts and fulfill increasing customer requirements."[38]

One innovative manufacturing method was the use of modular cells. The typical cell consisted of a group of machines set up in one central location to manufacture parts for the same component. The advantage of the cell technique had already proven itself. Products that once took six weeks

Woodward's Rockton, Illinois, plant

to build and ship now took only three days. Because of this sharply reduced lead time, plants could carry fewer pieces of finished products in inventory, another cost savings. In addition, production problems were easier to spot and solve because operators and supervisor were closer to each other and the actual manufacturing processes.

Cell manufacturing techniques also benefited Woodward's working members. At Stevens Point, for example, the Advanced Manufacturing Technology Group set up a cell to make pilot valve plates for aircraft controls. The process from start to finish took a raw casting through a computer numerical control (CNC) center, an electronic discharge machine (EDM), drilling, helicoil insertions, deburring, and inspection. Because the whole operation was centrally located in one cell, members had the opportunity to operate all the machines. This broadened their manufacturing skills, and it also enhanced their worth to the company, an important consideration at member evaluation time.

As Woodward enjoyed market success, introduced new products, and began employing new methods of operation, the company had little hint of the hardship it was about to face. Although prosperous activity would continue into the early 1990s, what was a vigorous marketplace would become much smaller, more complex, and highly competitive, placing much greater demands on Woodward. To succeed, the company would have to incorporate even greater degrees of change and do so in a very short time.

Positioning for the Future

A Time of Transition

1990-1995

December 1, 1994, marked the end of another period for Woodward Governor Company. On this day, Chairman of the Board and former Chief Executive Officer, Calvin C. Covert, died. Cal was active in the company until illness unexpectedly incapacitated him just a few short weeks before his death.

During the last few years of his life, Cal had to deal with problems beyond his control. These situations caused the company to take steps it never before considered.

The 1990s began with the world undergoing a time of change. During the 1980s, the Soviet Bloc began to disintegrate, creating an era of revived nationalism and promoting autonomy of member nations. A new Eastern Europe began to emerge. New governments formed, and East and West Germany rejoined as one nation. The reorganization of these nations created a degree of economic uncertainty. By the end of the decade, a worldwide recession began to emerge. This recession caused many companies to cancel or delay orders for applications using Woodward controls. As the decade ended, more than just the world's economy had begun to change.

With the fall of the Soviet Bloc, Western nations saw a significantly reduced military threat and began to scale back military spending. Many programs were canceled or reduced, and a number of these involved aircraft using Woodward controls.

In 1990, Iraq flexed its muscle and invaded Kuwait claiming territorial rights. The United Nations elected to confront Iraq and began a military build-up, which was based in Saudi Arabia. The increased political uncertainty caused by these moves, and the growing fear of international terrorism, added to the world's economic instability, particularly in the airline industry.

The USAF F-16XL with Woodward controls.

Small Aircraft Controls was located in a building leased from the Rockford School System, on 35th Street, Rockford.

With the new decade, improved control and control-system technology began to find wider applications, especially in the aircraft-control business. Engine makers and airframers were applying digital based systems on large numbers of all types and sizes of new aircraft. The purely hydromechanical, and even some older electromechanical, control concepts were becoming obsolete.

Aircraft, and aircraft engines, using Woodward controls were beginning to be replaced by new aircraft using newer engines. Even the highly successful Boeing 737-300, -400, and -500 aircraft, which used CFM56-3 engines with Woodward controls exclusively, were planned to be replaced with the newly designed Boeing 737-700, complete with a new engine and control system.

Although not as dramatically affected, Woodward's industrial markets also experienced a business slowdown. One cause of this situation may be traced to legislative bodies passing emission-control laws that forced engine manufacturers to change their method of controlling engines.

To meet emission standards over an engine's entire operating range, manufacturers began using electronic fuel injection systems. Since the sale of their engines heavily depended on the emission technology found in these new systems, many engine manufacturers decided to keep these systems within their own organizations. They made the decision for competitive reasons, and Woodward's growth suffered because these manufacturers were less willing to let independent companies provide controls or fuel systems. At the same time, the advent of the microprocessor allowed systems integration to increase. Integration gathered many previously independent functions into a one engine-control system and eliminated the need for the individual controls. Over all, Woodward faced increasingly fluid markets in a rapidly changing world.

As 1990 came to a close, the company still was enjoying improved sales and earnings. It was entering new markets and developing new products. The company's Aircraft Product Service Center was successful, and every indication pointed toward its continued growth. This success led toward organizing all the company's operations into focused business units.

During the year, the company established three business units. Large Aircraft Controls (LAC) and Aircraft Manufacturing business units were housed at the company's headquarters in Rockford, although Aircraft Manufacturing also had operations at the Stevens Point, Wisconsin, facility. The third business unit, Small Aircraft Controls (SAC), moved to a leased building on the south side of Rockford.

LAC had formed a partnership with General Electric to build part of the fuel control system for the new GE90 turbine. This was a turbine engine designed for larger aircraft proposed for future development, the first of which was the Boeing 777. SAC was involved with developing a control for the new FJ44, which was being built by Williams International. It was also working on a control for the new Pratt & Whitney Canada PW206 engine. In addition, it renewed development of a FADEC for aviation applications.

In Fort Collins, Colorado, three new business units were formed: Turbomachinery Controls, Engine Controls, and Manufacturing. All were located in the Fort Collins plant. These business units were developing and applying several new electronic control systems. Among these were the NetCon® digital control system for turbine applications and the complete locomotive control (CLC). Numerous other controls, in addition to a number of specialized valves were under development. These products were targeted toward specific markets.

Hydraulic Turbine Controls, another business unit, remained in Stevens Point. It had begun applying digital technology developed in Fort Collins in hydro applications and was finding success in doing so.

By this time, the company had begun to use cellular manufacturing techniques to replace batch manufacturing throughout its operations. The cell method helped cut lead times, and it gave members the ability to more fully participate in the company's operations.

The Woodward fuel control for the GE 90 turbine engine.

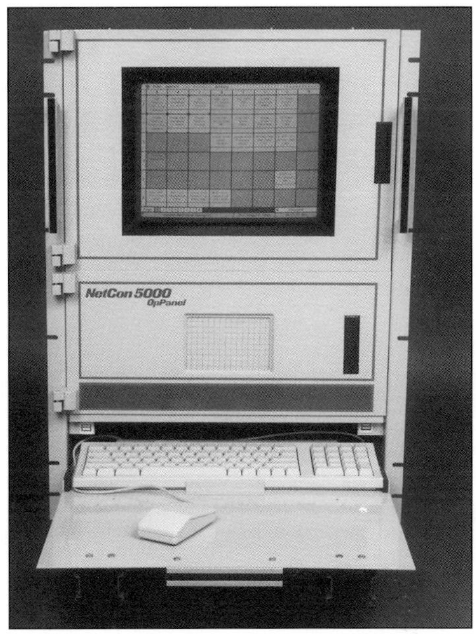

The NetCon DCS for turbine applications.

Outside the United States, Woodward had begun to expand its plant in Hoofddorp, the Netherlands, and was planning to open offices in Lyon, France, and in Singapore. Hoofddorp assumed responsibility for operations in Europe, the Middle East, and Africa. Also, the members in Campinas, Brazil, assumed responsibility for operations in South America.

Woodward plant in Hoofddorp, The Netherlands

Repairing an engine control at the APSC in Rockton, Illinois.

As the company ended the year, its 1990 Annual Report stated: "… we believe 1991 will be another rewarding, challenging year."

But change was in the air and beginning to make its mark on the company. The 1991 Annual Report began: "While 1991 fiscal year sales increased, year-end financial results were disappointing." During the year, sales of industrial controls increased by 18.4 percent, but sales of aircraft controls, which had been expected to increase, fell by 1.8 percent. Earnings dropped 17.5 percent.

Some of this drop in business related to the Gulf War, which began in February 1991. The United Nations invaded Southern Iraq, and drove the Iraqi troops from Kuwait. The war lasted only a few days, but it increased people's fear of traveling by air, further depleting airline profits. The airline industry found itself in a position where it suffered tremendous economic losses. The resultant airline industry economy had a rippling effect into many areas served by Woodward.

As airlines experienced losses in revenue, they began to cancel, or delay, purchases of new aircraft. These decisions had a direct effect on Woodward's aircraft control business. It also affected the Aircraft Product Service Center as controls sent in for repair dropped dramatically, particularly in the early part of the year.

However, work continued on aircraft-control development. Members demonstrated they were able to meet short production schedules. The fuel control for the FJ-44 engine was achieved in record time. SAC met all customer deadlines. SAC also met deadlines for the Pratt & Whitney

Canada PW206 control. Development of the new FADEC also was on schedule, as was LAC's GE90 effort.

Industrial Controls had a much stronger year in sales as more markets began to use Woodward products. Turbomachinery Controls introduced and applied the NetCon® 5000 F/T, a new fault tolerant control. Engine Controls shipped its first CLCs and formed an Aftermarket Parts Cell to better serve customers with spare parts.

The Hydraulic Turbine Controls business unit continued to apply digital technology, and with the NetCon 5000 control, it began to investigate the possibility of expanding from simple turbine control to complete plant control.

As part of the company's effort to make inroads into test equipment markets, it acquired Bauer Aerospace. Bauer was a company whose name was well recognized in the aircraft test industry and had experience making test and other types of equipment for aircraft applications.

Woodward also began constructing a building in Loveland, Colorado, to house the Turbomachinery Controls business unit. During the year, the company also completed construction of an addition to the Aircraft Product Service Center.

In June 1991, the board elected Carol J. Manning to the office of Corporate Secretary. Carol became the second woman to become a corporate officer, the first being Estelle Milne, who was elected to the position of Assistant Secretary in 1934. In September, Mary Barb Bittle retired from the board of directors.

The fuel control for the Williams International FJ44 engine

In November 1991, Mark Leum moved to the position of Vice Chairman of the Board, and John A. Halbrook was elected Woodward's President and Chief Operating Officer. John also became a member of the Board of Directors concurrent with his assuming the position of President.

Corporate Secretary Carol Manning

President and Chief Operating Officer John Halbrook

John began his Woodward career in 1984 as production manager for what was then the Engine & Turbine Controls Division. In October 1986 he became the division's manager of operations, and in January 1988 was appointed the division's general manager. In 1989 he was elected a vice president of the company. In August 1989, John relocated from Fort Collins, Colorado, to Rockford, Illinois, where he assumed the corporate responsibilities of Senior Vice President-Domestic Operations.

Woodward ended the year with a commitment to remain dedicated to customer service.

The 1992 Annual Report bluntly stated: "Although 1992 shipments were greater than 1991, they were less than our original expectations, and we were unable to achieve satisfactory financial results."

The commercial aviation industry continued in a slump. Military budgets continued to be cut. As a result, all areas of aircraft control sales were down. The one bright spot was the Aircraft Product Service Center, where the service of controls had picked up and remained strong.

Industrial Controls sales increased internationally, but declined domestically. Industries were expecting further economic downturns and were unwilling to invest money in new or upgraded equipment. Immediate opportunities were limited, but future opportunities remained hopeful with markets in Asia and along the Pacific Rim appearing to offer very good potential for future expansion.

To counter the decrease in business, the company took several drastic steps. It instituted a hiring freeze and reduced its work force by not replacing members lost through normal attrition. It moved members from indirect to direct labor positions and brought numerous contracted operations back into the shop.

In June, as business continued to look bleak, the company offered an early retirement package, and eighty-nine members took advantage of it. It also decreased the work force by ninety-three probationary members. During the year, total membership dropped 8.1 percent.

Customers were demanding shorter lead times. More cells and cellular evolutionary groups were formed in an attempt to meet these needs. Interdisciplinary work teams brought a sense of greater cooperation to the operation.

The successes in the Aircraft Controls business included the production of the new FJ-44 control. A prototype FADEC made a successful flight test in September. This project reached the flight-test stage within days of the original forecast. Engineers produced a working model of the GE90 control.

For existing customers, Aircraft Controls members drastically reduced cycle times for the CFM56-3 control. They also began work on a burner modulating valve for use on the CFM56-5B turbofan. The valve would help reduce exhaust emissions.

Rolls-Royce selected Woodward as a preferred supplier of hydromechanical controls. In addition, Woodward was selected by a new company, BMW Rolls-Royce to develop a control for a new family of turbofans, the BR700 series.

Bauer Aerospace suffered adverse effects from the economic problems of the airline industry. However, the Aircraft Product Service Center had a steady year and concentrated on shortening turnaround times to better meet customer needs.

Industrial Controls experienced increased sales over those of 1991. The business units had developed steady contacts with customers and end users. These contacts allowed the members to develop, or refine, controls to better meet customer needs. Turbomachinery Controls opened a regional office in Walnut Creek, California. Since its inauguration in the mid-1980s, the regional office program consistently yielded good results in putting a Woodward presence closer to strong customer bases and making customers aware of what the company had to offer them.

The burner modulating valve helps reduce exhaust emissions.

Woodward opened an office in Abu Dhabi, United Arab Emirates. The office supplemented those already in place in France, Germany, and Poland.

Retrofit applications constituted one of Industrial Controls' biggest opportunities. These opportunities are especially important since Woodward had the controls and resources to help clients with complete plant control. Automation was becoming very important in almost every industry served, and Woodward controls give the company the opportunity to meet these needs.

When 1992 ended, a broad, deep recession continued throughout the aviation industry. Aircraft Controls sales faced a not-too-promising future. Military spending was on the decrease. Fortunately, the prospects for Industrial Controls were brighter.

By the end of 1993, business had not improved. A sluggish global economy depressed world markets and the decline of Aircraft Controls shipments exceeded expectations. Industrial Controls shipments were less than predicted. In fact, the total value of shipments dropped 33.8 percent from 1992.

Woodward was experiencing increased pressure from customers to reduce prices and decrease development time. In addition, customers were demanding more sophisticated controls and control systems. Reduced fuel burn, reduced emissions, and increased control capabilities were major industry concerns. To meet this need, Woodward managers had to increase the funding of new product development.

In addition, the work force had become too large for the amount of business available. Particularly in the Aircraft Controls sector. To offset this, the company again offered the early retirement program and did not replace members lost through normal attrition. These actions alone were not enough to achieve management's goal, and in May Woodward had to permanently let some members go.

Woodward's Turbomachinery Controls in Loveland, Colorado

Before implementing the downsizing, the company evaluated each member to determine which positions were necessary and the importance of the member to overall operations. By the end of the year the total work force was cut to 3,264 members, a decrease of 10.1 percent when compared to 1992.

In the meantime, Canadair introduced its Regional Jet, an aircraft equipped with two GE TF35 turbofans, which were Woodward-controlled engines. The GE TF34 engine, and its control, were originally developed for military applications, but were now being used in civil aviation. In spite of the economic downturn, control development continued. The fuel control unit for the Pratt & Whitney Canada PW206 was ready, with the engine and airframe undergoing flight tests. Development of the GE90 control was on schedule, as was the fuel metering unit and overspeed splitter unit for the BMW Rolls-Royce BR710. Work continued on the burner modulating valve for the CFM56-5B turbine.

Parker Hannifin, a major supplier of nozzles to turbine manufacturers certified a nozzle cleaner made by Bauer Aerospace. Bauer used the same technology to develop cleaners for industrial use. The cleaners used environmentally acceptable solvents and would help users meet EPA standards. Bauer moved its business to larger quarters in late 1992. The larger building allowed it to expand its manufacturing and research and development areas.

Turbomachinery Controls also moved into the new Loveland, Colorado, plant. The move gave it and the Engine Controls business unit room to expand all operational areas. The NetCon 5000 system continued to be a leader in turbine control technology. During the year, Turbomachinery Controls introduced new operator interfaces and overspeed protection systems. The most important undertaking was the development of a new system to reduce turbine exhaust emissions.

Known as dry low NOx (DLN) or dry low emission (DLE) technology, the program put the business unit in the position of being one of the first suppliers of such a system. The project also called for the development of new fast-acting actuators and a variety of fuel shutoff valves.

Engine Controls developed several new digital systems that integrated several functions into one unit. These new products were helpful to customers repairing or upgrading existing systems. In addition to supplying upgraded technology, they simplified applications and saved space. The railroad industry began using Woodward's complete locomotive control and locomotive engine control systems.

Of particular significance was the inauguration of Central Distributor and Authorized Dealer networks throughout North America. When fully implemented, these networks were planned to increase sales of Woodward industrial controls in both the short and long term. Woodward also entered into a joint venture to open a plant in

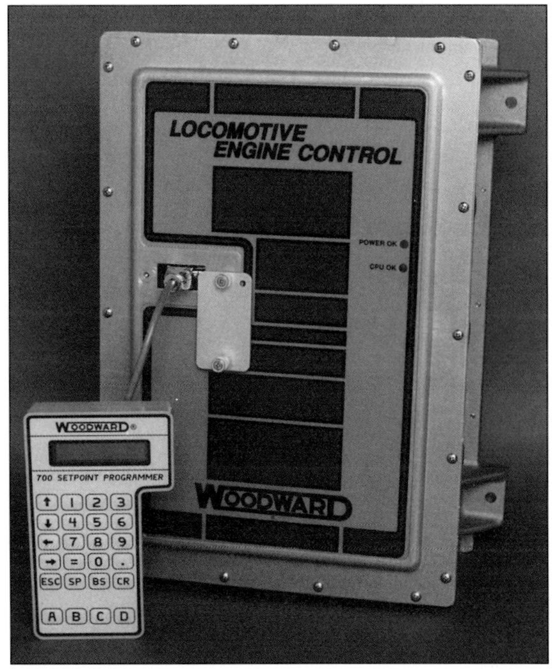

India. Fuel injection application remained a strong target for product development, and the company acquired a pump manufacturing plant in Aken, Germany, to gain the specialized technology needed for this market.

Hydraulic Turbine Controls inaugurated its first large-scale plant control system at the Philadelphia Electric Conowingo generating plant. Several other major application were completed this year. Overseas markets around the world held promise for a successful future.

Business did not improve in 1994. Industrial markets remained flat and aircraft controls markets continued in decline. However, future opportunities looked brighter.

The Woodward plant in India.

Total shipment values were 11.5 percent below those of the previous year. Although sales of Woodward aircraft controls continued to decline, the increase in industrial controls helped offset this decrease. Because the world's economy began to show signs of strengthening, there was more optimism for the future. However, the managers realized that whatever growth the company would experience would be slow. It would be a long time before Woodward experienced the financial success of the '80s. The world's markets had changed. Competition had changed. To survive in such a world required producing controls that delivered extremely high value for customer investment.

At this time, Woodward had too much manufacturing capacity for current and expected business activity. It took drastic measures to address this situation and make the decision to close two manufacturing facilities. Also, with the company's decision to pursue the development of fuel delivery systems, the Test Equipment business at the Bauer Aerospace plant no longer fit into the company's traditional or targeted business goals. Therefore, the company also decided to sell it.

One plant the company decided to close was that in Stevens Point, Wisconsin. Economically, the plant was too expensive to maintain in a highly competitive world. Managers decided to bring our aircraft parts manufacturing back to the Rockton and Rockford plants. They began to move the Hydraulic Turbine Controls business to Loveland, Colorado, where it was to become part of the Turbomachinery Controls business unit. The company also decided to close its leased facility in Rockford and bring its operations into the remaining Rockford and Rockton plants.

Many traditional internal programs also were cut to reduce costs. Several non-manufacturing functions along with a number of non-critical manufacturing operations were subcontracted. This action helped cut the work force and allowed members to concentrate on the critical phases of operations.

Managers reviewed the way the company conducted business, made changes where necessary, and set short- and long-term goals for the various business units. Aircraft Controls were pulled into one group, which began to operate out of the two plants in Illinois and plants overseas. The Large and Small Aircraft Controls business units, along with the Aircraft Product Service Center no longer existed.

After lengthy discussions of what direction the business should take, especially in regard to the aircraft business, the decision was made to work toward becoming a supplier not only of controls but of complete, fully integrated fuel delivery systems.

To complement this decision, Woodward acquired HSC Controls in Buffalo, New York. HSC manufactures electromechanical devices that are essential to build integrated control system, especially for aircraft engines. The company also purchased a small company in Kelbra, Germany, which manufactures fuel injection nozzles, essential for fuel injection applications. This company, together with the firm in Aken, Germany, became Woodward Governor Germany, GmbH.

In addition to restructuring business units, the company encouraged members to become self-empowered and assume more responsibility for their jobs. Teamwork was emphasized, and specialized teams were formed. Member training was conducted to help members develop team skills.

Just as some of these changes were beginning to get underway, Cal became ill and soon passed away. Leadership fell to John Halbrook, who as president, assumed responsibility of the company. At the plant meeting on January 15, 1995, John announced that on the previous day the board had elected him its chairman and the company's chief executive officer. With this action a new era began.

With John having the board's approval, he initiated additional steps to redefine the company. The anniversary date of the company's 125th year was approaching, and it was an appropriate time to reinvigorate the CorPar concept. It was time to reaffirm the partnership between all members as partners in the company. In addition, the dress code was relaxed, as were haircut and beard criteria. Coffee breaks were instituted.

Over time, many had come to think that longevity guaranteed a right to a job. John wanted to confront this false premise by reminding everyone that every day each member had to earn his or her right to participate in the company. The prevailing business climate mandated that simply showing up for work no longer was enough to achieve success. John wanted to redefine what every member could expect from the company, and what the company had a right to expect from every member. It was a large task to assume, but one that had to be achieved.

In 1995, as the company entered its 125th year of operation, business was beginning to look a little brighter. The world's economy was strengthening, and with it, both the airline and industrial markets were becoming more willing to invest in new or updated equipment.

Woodward engineers were busy working with customers to develop new controls. Many of the aircraft development programs were either in the last stages of flight test or had received certification. New projects were being inaugurated.

The dry low emissions (DLE) project had put Woodward at the forefront of DLE technology. The company had been chosen as a major DLE supplier to some of the world's largest turbine manufacturers. The Industrial Controls group were discovering new applications of existing products and introducing new controls designed for specialized markets. Woodward continued to open plants and offices throughout the world. One of the newest offices was in Beijing, China. The India operation had begun to ship products. Internally, Woodward was preparing for the future.

What does the future hold for Woodward Governor Company? If the past 125 years are an indication, the answer seems to be continued success in business. After all, Woodward's emphasis always has been based on delivering a good product at a fair price. Just as importantly, the company will continue to strive to be successful in an area more difficult to define—human resources. Maintaining this growth is essential because the company's products need to be built by a highly motivated work force—not just hired hands.

Woodward's legacy is solidly entrenched and supported by a constitution that stresses ethical business practices and the value of each individual. The Corporate Partnership continues to prosper. These values will be needed because the world is changing at a rate faster than ever before. For the company to succeed, we must hold onto our traditional values while changing antiquated traits into ones more viable to the new marketplace. The company will have to thrive on change in order to excel in its new environment.

From one man with an idea and the will to pursue it, emerged a company with over 3,000 members throughout the world. Woodward's past is being remembered in this book because what our men and women of Woodward did yesterday is relevant to what the men and women are doing today. In their turn, the current members of Woodward will become part of the company's history.

Notes

CHAPTER ONE. The Founder and His Family 1856-1872

1 *Rockford Republic,* March 25, 1919.

2 W.C.L.E. Ferslew, compiler, *Rock County Gazetteer, Directory and Business Advertiser for 1857-8* (Janesville: Ferslew and Co., 1857), p. 74. Fairbanks later was a clerk but never really prospered. He married but his wife died not long afterward. When he visited Maine in 1859, Amos's brother Henry saw him and described him as looking "miserable" and "dispeptic." He died in 1868.

3 "Old Settlers Corner: Amos W. Woodward," *Rockford Morning Star,* March 30, 1913.

4 Charles A. Church, *History of Rockford and Winnebago County,* (Rockford: W. P. Lamb, 1900), pp. 320-24.

5 "A Fairy Tale That Came True—The Woodward Governor," *Rockford Morning Star,* October 22, 1911.

6 Although none of Amos's letters survive, much can be inferred from the comments of his mother and other relatives regarding the contents of these lost letters. Copies of the surviving correspondence have been collected in the Woodward Governor Company Archives. The originals are in the possession of James Bittle, Amos's great-great grandson.

7 Maria to Amos, May 31, 1857.

8 Maria to Amos, December 6, 1857.

9 Henry to Amos, May 12, 1857.

10 Henry to Amos, June 8, 1957. Elwood, Kansas, appears to be the town mentioned.

11 Henry to Amos, July 2, 1857.

12 Nancy to Amos, October 23, 1857.

13 Ibid.

14 Henry to Amos, October 26, 1857. In a letter dated December 24, 1853, Henry disclosed his ambition to bring his boot business to the West.

15 *Rockford City Directory, and County Gazetteer, for 1869* (Rockford: Kauffman & Burch, 1869), p. 75; *First Baptist Church Record Book,* 1858, p. 217.

16 *First Baptist Church Record Book,* Regular Covenant Meeting, January 1, 1859, p. 8.

17 Ibid., p. 13. This resolution was adopted by the church on November 28, 1857.

18 *Rockford City Directory, and County Gazetteer, for 1869,* p. 72.

19 Nancy to Amos, July 27, 1858.

20 Henry to Amos, February 25, 1858; July 22, 1858; Priscilla to Amos, March 17, 1858; Nancy to Amos, July 27, 1858; December 28, 1858.

21 Henry to Amos, March 22, 1859; April 26, 1859.

22 Henry to Amos, April 26, 1859; Workers of the Writers' Program of the Work Projects Administration, compilers, *Rockford* [American Guide Series] (Rockford: Graphics Arts Corporation, 1941), p. 45. Hereafter cited as WPA, *Rockford.*

23 Franklin to Amos, April 27, 1859; Nancy to Amos, June 23, 1859; Henry to Amos, June 23, 1859.

24 *Williams' Rockford Directory, City Guide, and Business Mirror* (Rockford: J. W. Seccomb, 1859) p. 104; Henry to Amos, October 24, 1859; *The History of Winnebago County, Ill., Its Past and Present* (Chicago: H. F. Kett & Co., 1877), p. 419. There were three Mannys in Rockford. John H. Manny won the celebrated McCormick Case but died soon afterward in 1856. F. H. Manny, another relative, built the J. H. Manny reaper, and Amos Woodward worked for him beginning in 1859. John Pells Manny, a cousin of John H., invented his own reaper and mower, and N. C. Thompson built them. Amos Woodward also worked for Thompson later in the 1860s until he began his own business in 1872.

25 Priscilla to Amos, October 30, 1859; Henry to Amos, January 23, 1860.

26 Henry to Amos, April 4, 1860.

27 Nancy to Amos, May 17, 1861.

28 Nancy to Amos, May 17, 1861; Henry to Amos, September 22, 1861. At this point, there is a lapse in the family letters until 1864.

29 Henry to Amos, October 11, 1860; Charles A. Church, *Past and Present of the City of Rockford and Winnebago County, Illinois* (Chicago: S. J. Clarke, 1905), p. 724.

30 WPA, *Rockford,* pp. 40-42.

31 Sealy Family letters in possession of James Bittle.

32 Prudentia to Amos and Mary (Mate), May 14, 1867.

33 *Holland's Rockford City Directory for 1872-3* (Chicago: Western Publishing Co., 1872), p. 34; *Rockford City Directory and County Gazetteer, for 1869,* p. 97.

34 Church, *History of Rockford and Winnebago County,* p. 365; *Rockford City Directory, and County Gazetteer, for 1869,* pp. 13-14.

35 *The History of Winnebago County, Ill., Its Past and Present,* p. 419; James Lamont, "Reminiscences of the Water Power" in *Seventy Years of Progress, 1834-1904* (Rockford: *Register-Gazette,* 1904); WPA, *Rockford,* p. 32.

36 Ernest L. Bogart and Donald L. Kemmerer, *Economic History of the American People* (N.Y.: Longmans, Green and Co., 1948), pp. 382-83; Lamont, "Reminiscences of the Water Power."

37 Nancy to Amos, February 6, 1864.

38 *First Baptist Church Record Book,* pp. 40-87.

39 Amos to Mary (Mate), September 21, 1865; Sarah to Amos, June 18, 1866.

40 Grantee Book 7, Winnebago County Recorder's Office.

41 Nancy to Amos, January 13, 1868. Charles Fairbanks, contrary to company legend, was not related to the Fairbanks family of the Fairbanks-Morse Company.

42 Amos to Mary (Mate), July 5, 1868.

43 Amos to Mary (Mate), July 5, 1868; July 13, 1868; Mary (Mate) to Amos, July 19, 1868.

44 *Rockford City Directory, and County Gazetteer, for 1869,* p. 98; "A Fairy Tale That Came True—The Woodward Governor," *Rockford Morning Star,* October 22, 1911.

45 *Rockford Morning Star,* October 22, 1911; March 30, 1913.

CHAPTER TWO. The Infant Company 1872-1890

1 *Rockford Morning Star,* October 22, 1911.

2 Charles A. Church, *Past and Present of the City of Rockford and Winnebago County, Illinois* (Chicago, S. J. Clarke, 1905), p. 723; *Power Control,* August 1910, p. 4.

3 "Pressure Speed Governors" (list of patents ordered by Prime Mover Control Museum).

4 Gilbert Fite and Jim Reese, *An Economic History of the United States,* 3rd ed. (Boston: Houghton Mifflin, 1973), p. 290.

5 Malcolm Kier, *The Epic of Industry,* vol. V, *The Pageant of America* (New Haven: Yale University Press, 1926), p. 127.

6 Ross Robertson, *History of the American Economy,* 3rd ed. (New York: Harcourt Brace Jovanovich, 1973), p. 214.

7 Ibid., p. 338.

8 Alexis de Tocqueville, *Democracy in America.* Trans. Henry Reeve, ed. Andrew Hacker (New York: Washington Square Pocket Books, 1964), pp. 129-32.

9 WPA, *Rockford,* p. 60; James Lamont, "Reminiscences of the Water Power," in *Seventy Years of Progress, 1834-1904* (Rockford: *Register-Gazette,* 1904).

10 These jigsaws were manufactured by W.F. and John Barnes in Rockford.

11 WPA, *Rockford,* p. 130.

12 Ibid., pp. 49-50.

13 Grantee Book 12, Winnebago County Recorder's Office.

14 *Holland's Rockford City Directory for 1874-5* (Chicago: Western Publishing Co., 1874), pp. 190, 258.

15 *Brainerd's Rockford City Directory for 1876* (Rockford: A. Brainerd, 1876), p. 180.

16 Nancy to Amos, March 26, 1869; December 29, 1871; April 20, 1874. The family letters end in 1874. Whether no more were written until the early 1900s, when brother Henry's letters were kept, or whether letters between 1874 and 1906 were lost is not known.

17 *First Baptist Church Record Book,* pp. 88-96.

18 Ibid., pp. 107, 112, 114, 118, 124.

19 Ibid., pp. 129, 137, 138, 147, 151; Grantee Book 12, Winnebago County Recorder's Office.

20 *First Baptist Church Record Book,* pp. 220-389, passim.

21 Winfield Scott Downs, ed., *Encyclopedia of American Biography* (New York: The American Historical Co., 1942), p. 438; *Rockford Morning Star,* February 20, 1940.

22 *Rockford Register-Republic,* January 5, 1946; E. P. Bassett, *Bassett's Rockford City Directory for 1879* (Rockford: Abraham E. Smith, 1879), p. 129. The address at this time was listed as 601 Peach; later the number was changed to 603 Peach.

23 *Bassett's Rockford City Directory for 1879,* p. 129; *Matthews' Rockford City Directory for 1882-83* (Rockford: W. F. Matthews, 1882), p. 163.

24 *The History of Winnebago County, Ill., Its Past and Present* (Chicago: H. F. Kett & Co., 1877), pp. 416, 419-20.

25 Fite and Reese, *Economic History,* pp. 271-72.

26 Lamont, "Reminiscences of the Water Power;" Charles A. Church, *History of Rockford and Winnebago County* (Rockford: W. P. Lamb, 1900), p. 365; *Rockford Gazette,* September 15, 1884.

27 Fred C. Pierce, comp., *Picturesque and Descriptive History of the City of Rockford* (Rockford: Abraham E. Smith, 1887).

28 Ibid., p. 13.

29 *Rockford Weekly Gazette,* December 19, 1883, p. 2; Mortgage Records of A. W. Woodward.

30 *Eads' Illustrated History of Rockford,* (Chicago: A. B. Eads, 1884), p. 40.

31 WPA, *Rockford,* pp. 69-70.

32 Ibid.

33 Ibid., p. 71; Daniel Boorstin, *The Americans: The Democratic Experience* (New York: Vintage, 1973), pp. 534-36.

34 *Rockford Morning Star,* June 15, 1952 (Centennial Issue).

35 Martha Strolberg, *A Gentleman Named Elmer* (Rockford: Wilson-Hall Co., 1974), p. 12.

36 Letters they wrote to each other during this time are in possession of James Bittle; Woodward Governor Company Archives has copies. Although Mary Woodward was called "May" by family members, several of Elmer's letters addressed her as "Maimie" and "Mollie."

37 Elmer to Mary (May), February 6, 1887.

38 *Rockford Daily Gazette,* February 8, 1887.

39 Ibid., February 4, 1887; June 19, 1887.

40 Elmer to Mary (May), February 27, 1887.

41 Elmer to Mary (May), March 8, 1887.

42 Mary (May) to Elmer, July 22, 1887; Elmer to Mary (May), July 22, 1887.

43 Elmer to Mary (May), July 24, 1887.

44 Elmer to Mary (May), February 20, 1888.

45 *Rockford Daily Gazette,* September 11, 1886; October 28, 1886; November 1, 1886; December 27, 1886.

46 *Rockford Daily Gazette,* December 31, 1889; *Philippi's Rockford City Directory,* 1889-90, (Rockford: P. L. Philippi, 1889), p. 354.

47 *Serial Book 1 [1888-1910].* This is the earliest extant record of the sales of Woodward governors. The first entries are dated (beginning with April 13, 1888), but do not identify serial numbers. A complete dated record in serial number order begins with 570 (September 17, 1890) and ends with 2863 (December 28, 1910). The sequence continues with 2864 in *Serial Book 2 [1911-1924].*

48 *Rockford Daily Register,* December 31, 1889; January 24, 1890.

CHAPTER THREE. The Springboard to Modest Success 1890-1902

1 Elmer to Mary (May), March 30, 1890; Mary (May) to Elmer, March 31, 1890.

2 Eugene Browne and F. Ford Rowe, *Industrial and Picturesque Rockford* (Rockford: Forest City Publishing Co., 1891), p. 37.

3 Ibid., pp. 5, 7. Rockford's population was actually a little over 23,000 in 1890.

4 *Philippi's Rockford Blue Book* (Rockford: P. L. Philippi Co., 1890), pp. 262-63.

5 *Rockford Daily Register,* July 1, 1890; January 29, 1891; *Rockford Daily Register Gazette,* December 5, 1891.

6 Ibid., July 7, 1891; July 8, 1891.

7 *Serial Book 1 [1888-1910].*

8 Malcolm Kier, *The Epic of Industry,* vol. V, *The Pageant of America* (New Haven: Yale University Press, 1926), p. 121.

9 Catalogue and Price List, Woodward's New Friction Water Wheel Governor (Rockford: Monitor Publishing Co., 1892), pp. 3, 6.

10 Ibid., p. 8.

11 WPA, *Rockford,* p. 79.

12 *Power Control,* August 1910, p. 4; *Grantee Book 21,* Winnebago County Recorder's Office, May 1, 1893.

13 *Serial Book 1 [1888-1910].*

14 Richard Morris, ed., *Enyclopedia of American History* (N.Y.: Harper, 1961), p. 263; Gilbert Fite and Jim Reese, *An Economic History of the United States* (Boston: Houghton Mifflin, 1973), p. 272.

15 Fite and Reese, *Economic History,* p. 272; WPA, *Rockford,* p. 79. In Rockford the furniture cooperatives were the hardest hit since they lacked a cash reserve.

16 *Serial Book 1 [1888-1910].*

17 *Philippi's Rockford City Directory* (Rockford: [P. L.] Philippi Co., 1894).

18 WPA, *Rockford,* pp. 75-76.

19 U.S. Patents 583,527 and 608,245.

20 U.S. Patent 679,353 granted July 30, 1901.

21 *Serial Book 1 [1888-1910].*

22 Ibid.

CHAPTER FOUR. Incorporation to Construction 1902-1910

1 Walter LaFeber and Richard Polenberg, *The American Century* (New York: John Wiley, 1975), p. 28.

2 *All Roads Lead to Rockford* (Rockford Merchants' & Business Men's Association, ca. 1903); *Prime Mover Control,* December 1960, p. 3, for company statistics.

3 *All Roads Lead to Rockford,* pp. 2-8.

4 WPA, *Rockford,* pp. 81-82. In 1903 Britain set a 20-mile-per-hour speed limit for automobiles—a high figure for the time.

5 Ibid.

6 Fred Shannon, *America's Economic Growth,* 3rd ed. (New York: Macmillan, 1951), pp. 455, 615; WPA, *Rockford,* p. 83.

7 Shannon, *America's Economic Growth,* p. 489; WPA, Rockford, p. 83.

8 *Rockford Morning Star,* June 15, 1952 (Centennial Issue).

9 Irl Martin, "Woodward's Early Days," typescript, July 15, 1974.

10 Ibid.

11 Ibid.

12 *Rockford Daily Register-Gazette,* June 27, 1903.

13 Henry to Amos, September 11, 1907; October 4, 1907; December 19, 1907.

14 Henry to Amos, August 25, 1908.

15 Henry to Amos, March 11, 1909; Priscilla to Amos, December 26, 1910; February 1, 1911.

16 Henry to Amos, January 14, 1910; November 9, 1910.

17 Bert Woodward had worked for Savage & Love as a machinist in 1901 and was a foreman for Emerson Manufacturing Company in 1902. See *Philippi's Rockford City Directory for 1901-1902* and *The Republic's Rockford and Winnebago County Directory for 1902.*

18 Board of Directors, *Minutes,* May 14, 1902; June 28, 1902; July 15, 1902; July 15, 1903; J. Charles Young to Elmer, October 21, 1902; October 25, 1902.

19 Board of Directors, *Minutes,* June 28, 1902.

20 Statement to R. G. Dun & Co., January 1, 1908 lists the value of the company's assets as $77,548.66.

21 Photocopy of January 1903 article in unidentified Rockford newspaper. In actuality the company had not shipped machines to "almost every country on the globe," but the list was impressive.

22 *The Iron Age,* February 5, 1903; *American Electrician,* February 1903.

23 *Woodward Governor Company Catalogue,* (Rockford: Calbert-Wilson Co., c. 1904).

24 Ibid.

25 Ibid.

26 *Woodward Governor Company Catalog K,* c. 1910.

27 B. F. Groff to Woodward Governor Company, June 28, 1906; August 22, 1906.

28 Proposal to Fergus Falls, Minnesota, 1907; Proposal to Barton, Vermont, April 22, 1908.

29 Elmer Woodward to Mr. James Gable, York, Pennsylvania, July 3, 1908.

30 Proposal to Barton, Vermont, April 22, 1908.

31 *Serial Book 1* [1888-1910].

32 Ibid.

33 Ibid.

34 Board of Directors, *Minutes,* August 7, 1903.

35 Board of Directors, *Minutes,* 1904-1907. A fifty-five percent dividend would mean a cash payment during that fiscal year of $16,500, with Amos and Elmer getting at least $6,875 beyond their salaries.

36 Elmer to company members, December 24, 1910.

37 Grantee Book 34, Winnebago County Recorder's Office. These water rights (which had increased to 3,900 of 20,000 shares) were sold to Central Illinois Electric and Gas Company when Woodward made the decision in 1940 to build a new plant.

38 Board of Directors, *Minutes,* January 6, 1910; January 21, 1910.

39 *Power Control,* August 1910, pp. 4-5. Several different Mill Street address numbers were used for the Woodward plant. Although the 1910 and 1911 issues of the company's *Power Control* publication placed it at 201-211 Mill Street, city directories from 1909-1924 listed it as 240-250. Subsequent city directories gave the address as: 214-216 from 1925-1932; 216-218 from 1933-1939; 214-218 from 1940-1941.

40 Ibid., pp. 3, 5.

41 Ibid., p. 5.

CHAPTER FIVE. The Impact of World War I 1910-1919

1 WPA, *Rockford,* pp. 87-88.

2 Ibid., pp. 88-92.

3 Ibid., pp. 92-96.

4 Ibid., p. 97.

5 Ibid., pp. 102-3; Woodward Governor Company Federal Income Tax Return, 1917.

6 *Power Control,* November 1910, p. 4.

7 Robert Kelley obituary, *Rockford Register,* March 24, 1952; interview, Leo Powelson with George E. (Bud) Sorensen (George's son), October 1, 1985.

8 "Dear Gang," *Prime Mover Control,* November 1, 1948.

9 *Prime Mover Control,* March 1, 1947.

10 "A Fairy Tale That Came True: The Woodward Governor," *Rockford Morning Star,* October 22, 1911; *Bulletin L,* Woodward Governor Company, February 1912, pp. 3, 5, 7, 9, 10.

11 *Prime Mover Control,* June 1, 1949. This first V.R. hydraulic governor was number 2981, and it served the company and its successor for 38 years. In 1949, it was returned to Woodward to serve as a museum piece.

12 *Serial Book 2 [1911-1924].* The total sales figures are approximate since the company's serial books were not always in strict chronological order.

13 Ibid.

14 The forerunner of Dun & Bradstreet.

15 These figures were reached by adding the $30,000 capitalization to undivided profits in the statements to Dun & Co.

16 Board of Directors, *Minutes.*

17 Woodward Governor Company Federal Income Tax Returns; Board of Directors, Minutes.

18 Woodward Governor Company Federal Income Tax Returns, 1911-1915.

19 Interview, Leo Powelson with George E. (Bud) Sorensen (George's son), October 1, 1985; Board of Directors, Minutes, April 14, 1924.

20 *Rockford Register-Republic,* August 31, 1942.

21 Interview, Jean Ebling with Ward Ring, c. 1956.

22 Board of Directors, *Minutes,* June 13, 1919.

23 Interview, Jean Ebling with Ward Ring, c. 1956.

24 Woodward Governor Company Federal Income Tax Return, 1917; Board of Directors, *Minutes,* June 13, 1919.

25 Irl Martin, "Recollections," typescript, Summer 1974; "Dear Gang," *Prime Mover Control,* March 1962, p. 2.

26 "Dear Gang," *Prime Mover Control,* March 1962, p. 2.

27 "Dear Gang," *Prime Mover Control,* October 1962, p. 2.

28 Fred Humphrey got his M.D. degree and later was a Woodward plant physician and company director. His friendship with Irl lasted until Irl's death in 1976.

29 Interview, Dr. Fred Humphrey, May 30, 1986; "Dear Gang," *Prime Mover Control,* May 1962, p. 2.

30 Martin, "Recollections."

31 Irl to Dorothy, December 22, 1918. Irl Martin's letters to Dorothy are in possession of his grandson, James Bittle; copies in Woodward Governor Company Archives.

32 Ibid., July 30, 1918; October 5, 1918.

33 Ibid., November 13, 1918.

34 Ibid., December 10, 1918; December 22, 1918; December 25, 1918.

35 *Rockford Republic,* March 25, 1919.

CHAPTER SIX. The Prosperous 1920s 1920-1929

1 Irl Martin, "Recollections," typescript, Summer 1974. Elmer's son, Walter, was involved in the company as an officer in the 1920s, but by the 1930s he was at best a part-time officer and definitely not a potential president.

2 Ibid., p. 3.

3 Ibid.

4 Interview, Luella Lindholm, May 28, 1986.

5 Ibid.

6 This situation would last until the mid-1980s.

7 Gilbert Fite and Jim Reese, *An Economic History of the United States,* 3rd ed. (Boston: Houghton Mifflin, 1973) pp. 489-90.

8 WPA, *Rockford,* pp. 98-99; *Milestones Along the Trail of Progress: The Centennial of Rockford,* unpaged pamphlet, ca. 1934. From 1924 to 1940, Camp Grant was reserved for the Illinois National Guard, but during World War II it was again taken over by the federal government and expanded.

9 WPA, *Rockford,* p. 103.

10 Ibid., pp. 102-4.

11 Ibid., p. 104; *Milestones Along the Trail of Progress.*

12 WPA, *Rockford,* pp. 106-7; Gail Ravitts, "The Perils of the Greater Rockford," *Rockford Magazine,* February 1987. The plane was lifted off the ice cap by helicopter in 1968 and returned to Rockford in 1969. It was rebuilt and displayed at Rockford Museum Center.

13 Irl Martin, "Assets and Liabilities," typescript, ca. 1972; "Dear Gang," *Prime Mover Control,* December 1960, pp. 2-3. Irl Martin started at 60¢ per hour in 1921. He was time clock number 32 in 1921 and, despite some fluctuations, the company still had only 32 members in 1933.

14 *Water Power Control,* January 1929, p. 3; *Rockford Register Republic,* January 1, 1941; R. J. Kelley to D. E. Benjamin, June 20, 1930.

15 Interview, Jean Ebling with Ward Ring, ca. 1956; "Dear Gang," *Prime Mover Control,* August 1, 1949; Martin, "Recollections."

16 Interview, Jean Ebling with Ward Ring, ca. 1956.

17 The provisions of Amos's will are not known; his Estate File 6888 is missing from the Winnebago County Circuit Clerk's Office. Mary (Mate) died without a will, and Elmer administered her estate. Its value of less than $40,000 included 45 shares of the capital stock of the Woodward Governor Company valued at $29,250. Thus, each share was worth $650. See Mary A. Woodward Estate File 7942, Winnebago County Circuit Clerk's Office.

18 Board of Directors, *Minutes,* November 18, 1921; Mary A. Woodward Estate File 7942, Winnebago County Circuit Clerk's Office; Statement of Woodward Governor Company to R. G. Dun & Company, January 1, 1930.

19 Board of Directors, *Minutes,* February 21, 1928.

20 *Woodward Water Wheel Governors: Net Price List Number 10,* ca. 1925.

21 *Electrical World,* August 1922 (as excerpted in *Rockford-Register Gazette,* August 18, 1922).

22 These figures are yearly approximations based on *Serial Book 2 (1911-1924).*

23 Ibid.

24 Statements to R. G. Dun & Co.

25 Board of Directors, *Minutes,* 1920-1929.

26 Board of Directors, *Minutes,* December 22, 1922; *Prime Mover Control,* December 1960, p. 3.

27 "Dear Gang," *Prime Mover Control,* October 1, 1949; Irl Martin, "How Governor School Got Started," typescript, July 15, 1974.

28 Raymer B. Olmstead, Seward, Nebraska, to George Sorensen, March 4, 1928.

29 *Water Power Control,* April 1928.

30 Ibid., May 1929.

31 Ibid., May 1930.

CHAPTER SEVEN. From Depression to Expansion 1930-1939

1 William Manchester, *The Glory and the Dream* (New York: Bantam, 1975) pp. 33-36; *Historical Statistics of the United States* (Washington, D.C.: U.S. Dept. of Commerce, 1975), p. 126.

2 WPA, *Rockford,* p. 110.

3 Ibid., pp. 109-10.

4 Ibid., p. 110.

5 Ibid., pp. 112-16.

6 These figures are based on the company's original federal income tax returns. After later audits some of these figures were changed slightly. All figures are to the nearest thousand.

7 *Water Power Control,* May 1930, p. 4.

8 *Water Power Control,* May 1930, pp. 2-3; April 1931; May 1931.

9 Irl Martin, "Recollections," typescript, Summer 1974; *Water Power Control,* October 1928; November 1928; February 1931.

10 Note dated May 31, 1930.

11 Board of Directors, *Minutes,* July 12, 1932; February 20, 1934. Robert Kelley remained a vice president, but was no longer a director.

12 Eventually creditors got back 72¢ on the dollar, and as of 1941 the building was for sale, which meant creditors might get more. (See WPA, *Rockford,* p. 111.)

13 "Dear Gang," *Prime Mover Control,* August 1, 1949, December 1, 1961, June 1975.

14 Ibid.; interview, Luella Lindholm, May 28, 1986.

15 *Water Power Control,* October 1932, p. 4; "Dear Gang," *Prime Mover Control,* August 1949; December 1961; June 1975; Woodward Governor Company, Federal Income Tax Returns, 1930-1937; Board of Directors, *Minutes,* 1930s.

16 *Water Power Control,* June 1932, pp. 1-2; October 1932, pp. 2-3; interview, Leo Powelson with George E. (Bud) Sorensen, October 1, 1985; interview, Norm Noling, July 17, 1986; interview, Urban Floor, June 25, 1986. Al Kalin would remain a valuable member of the organization until the late 1930s. While Elmer was fond of Kalin and impressed with his brilliance, Kalin eventually alienated both Elmer and Irl. Shortly after Forrest Drake began his engineering career at Woodward in 1938, Kalin was terminated.

17 *Water Power Control,* April 1933, pp. 1-3.

18 Irl Martin, "Recollections;" "Dear Gang," *Prime Mover Control,* November 1949, p. 2. Some company people thought the IC stood for Irl C. Martin. Elmer's first IC was functional but not symmetrical. Irl wanted him to make it more attractive. Although his response was "beauty is as beauty does—this governor works perfectly," he agreed and the IC was adapted and made more attractive.

19 Ibid. Urban Floor thought he SI governor was "a dog," but it was easy to work on.

20 "Dear Gang," *Prime Mover Control,* November 1949; Daniel Cox to Irl Martin, October 30, 1934.

21 Rex Wadman to Irl Martin, October 2, 1934; Daniel Cox to Irl Martin, October 30, 1934. This development probably was related to the SI governor project at Winton Engine. George Codrington was Winton's general manager and owned the "All Alone."

22 *Serial Book 1 [1924-1940];* E. E. Woodward, "Report," handwritten, January 30, 1934. This *"Serial Book 1,"* which includes entries 5438 (November 28, 1924) through 14,607 (October 28, 1940) is actually the third of the company's governor sales record books in serial number order. It is, however, the first of a numbered multi-volume set.

23 "Dear Gang," *Prime Mover Control,* November 1949; November 1957; October 1969.

24 Ibid.

25 Interview, Norm Noling, July 17, 1986.

26 *Prime Mover Control,* June 1935, pp. 1-3; January 1937, p. 2; interview, Norm Noling, July 17, 1986.

27 Interview, Urban Floor, June 25, 1986; interview, Art Nielsen, June 20, 1986.

28 Interview, Urban Floor, June 25, 1986.

29 Irl Martin to R. J. Kelley, March 2, 1939.

30 *Prime Mover Control,* October 1937, pp. 1-3.

31 "Program for A.S.M.E. National Spring Meeting," (Los Angeles, March 23-25, 1938); Elmer's handwritten notes, for his presentation titled "Modern Turbine Governing."

32 *Rockford Morning Star,* November 28, 1938; October 9, 1938.

33 *Prime Mover Control,* November 1938.

34 *Rockford Morning Star,* January 26, 1939.

35 Ibid., February 20, 1940.

36 Interview, Edward (Ned) Garst, June 27, 1986.

37 Interview, Norm Noling, July 17, 1986.

38 Interview, Edward (Ned) Garst, June 27, 1986.

39 Interview, Leo Powelson, June 10, 1986.

40 Ibid.

41 Ibid.

42 Ibid.

CHAPTER EIGHT. The Impact of World War II 1940-1945

1 "Dear Gang," *Prime Mover Control,* May 1950.

2 "Dear Gang," *Prime Mover Control,* June 1950; Board of Directors, *Minutes,* November 1, 1939.

3 Board of Directors, *Minutes,* November 1, 1939.

4 Board of Directors, *Minutes,* November 11, 1939.

5 Elmer to Irl, February 8, 1940.

6 Ibid.

7 Elmer to Irl, undated letter (shortly after February 8, 1940).

8 Interview, Leo Powelson, 1987. Eugene Abegg later became a director of the company and remained on the board for decades.

9 Day & Zimmerman, Inc., *Report No. 3563,* June 5, 1940, p. 15; *Prospectus,* Woodward Governor Company, July 30, 1940, p. 15.

10 *Report No. 3563,* June 5, 1940, pp. 44-45.

11 *Prospectus,* July 30, 1940; "Dear Gang," *Prime Mover Control,* February 1959.

12 Board of Directors, *Minutes,* August 12, 1940. The 25¢ per quarter dividend was the standard throughout World War II.

13 Elmer Woodward to Members of the Woodward Governor Company, September 4, 1940.

14 *Rockford Morning Star,* November 12, 1940; *Annual Report,* September 30, 1940. In 1940, the company's fiscal year was changed from the calendar year to an October 1-September 30 schedule. Thus, the 1940 annual report covers only the first nine months of that year.

15 Board of Directors, *Minutes,* December 5, 1940.

16 *Rockford Register-Republic,* January 1, 1941; January 2, 1941; Resolution by Rockford Rotary Club, January 9, 1941.

17 Plant Meeting Tape 16, May 11, 1945. Irl began the practice of audiotaping plant meetings in 1942.

18 Plant Meeting Tape 2, Side A, May 1942.

19 Interview, Cal Covert, May 21, 1987. Apparently Irl was convinced that all of the tension associated with the company's increased activities precipitated Elmer's death, and he blamed himself.

20 Elmer E. Woodward Estate File 15815, Winnebago County Circuit Clerk's Office; interview, Cal Covert, May 21, 1987; Board of Directors, *Minutes,* 1940-1945.

21 *Introduction Manual for Applicants,* Woodward Governor Company, December 16, 1940.

22 Ibid.

23 Ibid.

24 *Annual Report,* September 30, 1941.

25 Interview, Edward (Ned) Garst, June 27, 1986; interview, Urban Floor, June 25, 1986.

26 Irl Martin, "News Bulletin," August 1941; *Annual Report,* September 30, 1941. The sun room was never very popular despite Irl's encouragement, and it was removed after a few years of modest use. Bowling alleys were added in 1947, but later removed when more space was needed.

27 UAW, "To the Employees of the Woodward Governor Company," 1941; "Dear Gang," *Prime Mover Control,* April 1, 1950.

28 Board of Directors, *Minutes,* January 31, 1941; *Annual Report,* September 30, 1942.

29 Board of Directors, *Minutes,* April 15, 1941; August 6, 1941.

30 Ibid., September 29, 1941.

31 Ibid., October 31, 1941; December 4, 1941.

32 *Rockford Register-Republic,* December 1, 1941.

33 Plant Meeting Tape 1, Side B, March 1942.

34 Plant Meeting Tapes 1-17, 1942-45.

35 Board of Directors, *Minutes,* May 9, 1942.

36 *Prime Mover Control,* 1942-1945.

37 "Dear Gang," *Prime Mover Control,* June 1, 1944; Irl Martin, "Recollections," typescript, Summer 1974.

38 G. Forrest Drake, "Governing Handbook," unpublished manuscript, 1971.

39 Board of Directors, *Minutes,* June 2, 1942; *Rockford Register-Republic,* May 28, 1942.

40 *Rockford Register-Republic,* October 8, 1942; *Prime Mover Control,* December 1942.

41 Plant Meeting Tape 4, Side A, September 4, 1942.

42 Plant Meeting Tape 7, Side A, November 2, 1942.

43 Plant Meeting Tape 7, Side B, December 7, 1942; *Prime Mover Control,* December 1942.

44 Board of Directors, *Minutes,* May 7, 1943; July 5, 1943. The company also trademarked its distinctive Woodward logo in 1943. It had been used since August 1917.

45 Plant Meeting Tape 9, Side A, February 1, 1943. Irl disliked Franklin Roosevelt's New Deal as much as he admired Theodore Roosevelt's Square Deal.

46 Plant Meeting Tape 9, Side B, November 17, 1943.

47 G. Forrest Drake, "The Development of Gas Turbine Controls," typescript of A.S.M.E. paper, ca. 1952.

48 Interview, Edward (Ned) Garst, June 27, 1986.

49 Interview, Urban Floor, June 25, 1986; interview, Edward (Ned) Garst, June 27, 1986.

50 Ibid.

51 G. Forrest Drake to Irl Martin and Roy Estell, memo (stamped "SECRET") on "Jet Engine Governor Development," February 23, 1945; G. Forrest Drake, "The Development of Gas Turbine Controls," 1952.

52 Drake to Martin and Estell, "Jet Engine Governor Development," February 23, 1945.

53 Plant Meeting Tape 10, Side A, undated (but clearly late February or early March 1944).

54 Plant Meeting Tape 10, Side B, April 19, 1944; Plant Meeting Tape 11, Side A, May 4, 1944.

55 "Dear Gang," *Prime Mover Control,* June 1, 1944.

56 Board of Directors, *Minutes,* January 10, 1944; "Dear Gang," *Prime Mover Control,* July 15, 1944.

57 "Dear Gang," *Prime Mover Control,* September 15, 1944; October 1, 1944.

58 M. S. Rukeyser, "Attractive Quarters Production Stimulus," *Los Angeles Examiner,* October 20, 1944

CHAPTER NINE. The Formation of the Corporate Partnership 1945-1954

1 "Dear Gang," *Prime Mover Control,* September 1944.

2 *Rockford Morning Star,* August 21, 1945.

3 *Rockford Morning Star,* September 11, 1945.

4 *Prime Mover Control,* 1942-1945. According to Irl, 86% of Woodward's service personnel became commissioned or non-commissioned officers.

5 William Manchester, *The Glory and The Dream* (New York: Bantam, 1975); pp. 388, 396.

6 Ibid., p. 419.

7 "Dear Gang," *Prime Mover Control,* July 15, 1945.

8 Plant Meeting Tape 16, Side A, May 11, 1945.

9 Irl Martin, "Encouraging Employees To Think For Themselves," *Illinois Journal of Commerce,* January 1944, p. 24.

10 Plant Meeting Tape 18, September 28, 1945. Although Irl was still referring to "bonuses" in 1945, this terminology was eventually replaced in official usage by the more appropriate "profit sharing."

11 Irl Martin, "A Business Philosophy," 1947. This typescript of a speech Irl presented to several business organizations was revised and published in 1950, as *Woodward Bulletin 25005.*

12 Ibid.

13 "Dear Gang," *Prime Mover Control,* July 1, 1945.

14 Martin, "A Business Philosophy," p. 11.

15 Ibid.

16 *Prime Mover Control,* June 1951.

17 Martin, "A Business Philosophy," p. 16.

18 *Annual Report,* 1946-1954.

19 *Prime Mover Control,* January 1955.

20 Norm Noling to General Electric, December 24, 1954.

21 *Prime Mover Control,* November 1948; January 1949; January 1950.

22 *Prime Mover Control,* January 1951.

23 *Prime Mover Control,* January 1952; January 1953.

24 *Prime Mover Control,* January 1953.

25 *Prime Mover Control,* January 1954; January 1955.

26 *Prime Mover Control,* January 1947.

27 *Prime Mover Control,* January 1948.

28 *Prime Mover Control,* January 1949; January 1950; January 1951.

29 *Prime Mover Control,* January 1951.

30 *Prime Mover Control,* January 1952; January 1953.

31 *Prime Mover Control,* January 1954; January 1955.

32 *Prime Mover Control,* January 1955.

33 Robert Phillips, unpublished draft of magazine article, 1950, p. 15. In a letter dated May 31, 1950, Phillips expressed concern about reaction to his draft to Harold A. Hopkins, of Woodward Governor Company.

34 "Dear Gang," *Prime Mover Control,* September 1950.

35 Luke Avery, "South American Diary," *Prime Mover Control,* September 1950.

36 Irl's analogy was based on personal experience with a recurring case of "turista."

37 "Dear Gang," *Prime Mover Control,* July 1953.

38 *Prime Mover Control,* September 1949; February 1951. Woodward mathematician, Dr. Rufus Oldenburger, had his book *Mathematical Engineering Analysis* published by MacMillan in 1951. It applied mathematics and physics to engineering. Oldenburger was a force in the development of the analog computer as well.

39 "Dear Gang," *Prime Mover Control,* April 1952; March 1953. Another method was tried in 1948, but it was not equitable. It was dropped after that year.

CHAPTER TEN. Domestic and Foreign Expansion 1955-1965

1 *Annual Report,* 1959; 1969.

2 C. Hal Nelson, ed. *Sinnissippi Saga* (Mendota, Illinois: Wayside Press, 1968) pp. 518-23.

3 Ibid., p. 522.

4 Ibid., p. 526.

5 Ibid., pp. 522-29. Several other Woodward members were elected to various local and regional offices.

6 Information on Billy Bittle from interviews with: Mary Barb Bittle; Edward (Ned) Garst; Urban Floor; Leo Powelson; Cal Covert; Luella Lindholm; Rex Hunt; Norm Noling.

7 Interview, Cal Covert; interview, Norm Noling; interview, Rex Hunt; interview, Leo Powelson; *Prime Mover Control,* March 1964.

8 *Prime Mover Control,* March 1964.

9 Ibid.

10 "Dear Gang," *Prime Mover Control,* May 1958.

11 *Rockford Morning Star,* December 28, 1954; interview, Dr. Fred Humphrey, May 30, 1986. Dr. Humphrey served on the company's Board of Directors from 1961 until 1971, when he was named emeritus director.

12 *Rockford Morning Star,* December 28, 1954; *Fort Collins Coloradoan,* March 4, 1955.

13 "Dear Gang," *Prime Mover Control,* May 1958; *Prime Mover Control Supplement,* December 1956.

14 Interview, Jack McMillan, May 30, 1986; interview, Cy Plessinger, May 30, 1986; interview, Fred Lowery, May 30, 1986.

15 "Woodward Governor Company, Fort Collins—History," updated July 1982, in *Worldwide Plant Histories* (loose-leaf binder).

16 "Dear Gang," *Prime Mover Control,* February 1965.

17 Ibid., October 1955.

18 Ibid., November 1955.

19 "Woodward Governor Nederland B.V.—History," updated February 1985, in *Worldwide Plant Histories* (loose-leaf binder).

20 "Dear Gang," *Prime Mover Control,* April 1958; interview, Woody Carson, September 15, 1986.

21 Ibid.

22 "Dear Gang," *Prime Mover Control,* April 1958.

23 "Woodward Governor (U.K.) Limited—History," updated February 1985, in *Worldwide Plant Histories* (loose-leaf binder).

24 "Woodward Governor (Japan) Ltd.—History," updated May 1984, in *Worldwide Plant Histories* (loose-leaf binder).

25 "Dear Gang," *Prime Mover Control,* September 1965.

26 *Prime Mover Control,* January 1957, p. 10.

27 Interview, Urban Floor, June 25, 1986; *Prime Mover Control Supplement,* March 1957.

28 Interview, Rex Hunt, July 23, 1986.

29 Interview, Jim Sundstedt, May 30, 1986.

30 *Prime Mover Control,* June 1959.

31 *Prime Mover Control,* August 1959.

32 *Prime Mover Control,* September 1961.

33 *Prime Mover Control,* February 1965.

34 *Prime Mover Control,* December 1964, p. 6.

35 Interview, Leo Powelson, July 8, 1987; *Prime Mover Control,* January 1958.

36 *Prime Mover Control,* January 1959; January 1960.

37 *Prime Mover Control,* January 1961; January 1962; January 1963; December 1963; December 1964

38 *Prime Mover Control,* January 1958; January 1959; January 1960.

39 *Prime Mover Control,* January 1961; January 1962; January 1963; December 1963; December 1964.

40 *Prime Mover Control,* June 1960; July 1961; September 1964; January 1965; June 1965.

41 *Annual Report,* 1965.

42 Ibid.

43 "Dear Gang," *Prime Mover Control,* February 1960; *Prime Mover Control Supplement,* July 1960.

44 *Prime Mover Control Supplement,* June 1961.

45 Interview, Cal Covert, May 21, 1987; *Prime Mover Control,* March 1965.

46 Interview, Cal Covert, May 21, 1987; "Dear Gang," *Prime Mover Control,* April 1965.

47 Board of Directors, *Minutes,* December 28, 1965.

48 "Dear Gang," *Prime Mover Control,* September 1963.

CHAPTER ELEVEN. The End of an Era But Not of an Idea 1966-1976

1 William H. Chafe, *The Unfinished Journey: America Since World War II* (New York: Oxford University Press, 1986) p. 401.

2 *Prime Mover Control,* February 1966; interview, Cal Covert, May 27, 1987; interview, Leo Powelson, July 8, 1987.

3 *Prime Mover Control,* May 1957; *Prime Mover Control Supplement,* May 1957.

4 "Dear Gang," *Prime Mover Control,* March 1967.

5 Interview, Vern Cassens, July 21, 1987; interview, Phil Turner, July 27, 1987.

6 "Summary of Operations—Ten Year Record," *Annual Report,* 1976, p. 9.

7 Board of Directors, *Minutes,* 1966-1976.

8 Ibid.

9 Ibid.

10 *Annual Report,* 1966; 1976. Aircraft Controls did not return to over 50% of sales until 1985.

11 Ibid.

12 Board of Directors, *Minutes,* December 28, 1966.

13 Ibid., August 5, 1966.

14 Ibid.

15 Ibid., March 3, 1967; August 9, 1967.

16 "Dear Gang," *Prime Mover Control,* July 1972.

17 *Rockford Register-Republic,* February 15, 1968 (Sesquicentennial Edition), sect. 3, p. 9.

18 "Dear Gang," *Prime Mover Control,* July 1969.

19 *Prime Mover Control,* January 1971. This *Prime Mover Control* article erroneously reported the new name as the "Irl C. Martin Academy of Industrial *Skills.*"

20 Interview, Phil Turner, July 27, 1987. Phil received $500 a year from Irl when he attended Northwestern University.

21 "Dear Gang," *Prime Mover Control,* July 1972.

22 Board of Directors, *Minutes,* September 20, 1967.

23 Board of Directors, *Minutes,* 1966-1970; *Annual Report,* 1966-1970.

24 Board of Directors, *Minutes,* August 13, 1968.

25 Ibid, December 19, 1969.

26 Board of Directors, *Minutes,* September 19, 1968; March 3, 1969; August 13, 1969; December 19, 1969; September 1, 1970. The company concluded on September 1, 1970, after a long study, that there was not yet enough interest on the part of worker members for a stock purchase plan to be established.

27 Social Concerns Committee, *Minutes,* July 31, 1969.

28 News Release, Woodward Governor Company, May 31, 1970.

29 "Dear Gang," *Prime Mover Control,* May 1970; Board of Directors, *Minutes,* April 3, 1970. No spring profit sharing was paid in 1970.

30 Board of Directors, *Minutes,* June 14, 1971; "Dear Gang," *Prime Mover Control,* May 1971.

31 Board of Directors, *Minutes,* June 14, 1971.

32 "Dear Gang," *Prime Mover Control,* August 1971.

33 Board of Directors, *Minutes,* January 5, 1972.

34 Board of Directors, *Minutes,* February 2, 1972; September 20, 1972; "Dear Gang," *Prime Mover Control,* April 1972.

35 Board of Directors, *Minutes,* July 12, 1972.

36 Annual Report, 1975.

37 "Dear Gang," *Prime Mover Control,* May 1973; Board of Directors, *Minutes,* March 21, 1973.

38 Board of Directors, *Minutes,* November 14, 1973.

39 Board of Directors, *Minutes,* March 20, 1974; May 21, 1974; July 17, 1974. The company eventually paid profit sharing of over 40% in 1974.

40 *Prime Mover Control,* July 1973; Board of Directors, *Minutes,* March 20, 1974; December 11, 1974; January 8, 1975; January 7, 1976.

41 "Dear Gang," *Prime Mover Control,* May 1975; *Annual Report,* 1976; *Prime Mover Control,* December 1976.

42 Board of Directors, *Minutes,* December 11, 1974; September 23, 1975.

43 Board of Directors, *Minutes,* September 23, 1975.

44 Board of Directors, *Minutes,* January 7, 1976; January 15, 1976.

45 Irl Martin to Board of Directors, March 24, 1976.

46 Board of Directors, *Minutes,* March 24, 1976.

47 Ibid.

48 Reverend Ralph Fry, April 23, 1976.

49 Radio Station WYFE Editorial, April 28, 1976.

50 *Rockford Register Star,* May 20, 1976; interview, Cal Covert.

CHAPTER TWELVE. A Time of Growth 1976-1989

1 Interview, Cal Covert, May 21, 1987.

2 William H. Chafe, *The Unfinished Journey: America Since World War II* (New York: Oxford University Press, 1986), p. 447.

3 Interview, Cal Covert, May 21, 1987; May 27, 1987.

4 Ibid.

5 Ibid.

6 *Rockford Magazine,* January 1989.

7 Board of Directors, *Minutes,* September 21, 1982.

8 Board of Directors, *Minutes,* March 5, 1983; June 29, 1983.

9 Board of Directors, *Minutes,* June 24, 1987.

10 Board of Directors, *Minutes,* September 21, 1976.

11 Interview, Vern Cassens, July 21, 1987.

12 Board of Directors, *Minutes,* June 25, 1980; January 7, 1981;

13 *Annual Report,* 1980, 1981.

14 Proxy Statement for Annual Meeting of Shareholders, January 11, 1984.

15 Robert McGough, "How to Win the Class Struggle," *Forbes,* November 3, 1986, p. 154.

16 Paul Weingarten, "Woodward's Way," *Chicago Tribune Magazine,* July 29, 1984; interview, Cal Covert, July 16, 1987.

17 Board of Directors, *Minutes,* January 7, 1981.

18 Board of Directors, *Minutes,* 1972-1982; *Annual Report,* 1972-1982. The 1972 per share figure is adjusted to reflect the 4-for-1 stock split in January 1981.

19 *Annual Report,* 1989.

20 *Prime Mover Control,* December 1977.

21 *Prime Mover Control,* December 1977; December 1982.

22 *Prime Mover Control,* December 1978; December 1979.

23 *Prime Mover Control,* December 1981; December 1982; December 1983; December 1986.

24 *Prime Mover Control,* December 1986.

25 *Annual Report,* 1981.

26 *Annual Report,* 1986.

27 *Prime Mover Control,* December 1985; December 1986.

28 Board of Directors, *Minutes* 1976-1987; *Wisconsin Business,* April 1986.

29 *Annual Report,* 1988.

30 Cal Covert, address, Annual Meeting of Shareholders, 1989.

31 Ibid.

32 Ibid.

33 Ibid.

34 *Prime Mover Control,* December 1988, pp. 11-12.

35 *Prime Mover Control,* December 1988, p. 4.

36 *Prime Mover Control,* December 1988, pp. 4-7.

37 Ibid., pp. 7-8.

38 Ibid., p. 14.

CHAPTER THIRTEEN. Positioning for the Future 1990-1995

Woodward Annual Reports 1990-1994